Maths

OUNDATION | MODULAR

Homework Book

Claire Turpin

About this book

This book provides extra exercises for the topics covered in the Foundation Student Book. There are four or five exercises for each Student Book chapter. In each, the first exercise (HW1) reviews topics from previous chapters in that Unit; the final exercise reviews the chapter and includes some exam-style questions. The remaining two or three exercises for each chapter give extra practice in the key topics.

Contents

OXFORD
UNIVERSITY PRESS

Great Clarendon Street, Oxford OX2 6DP

Oxford University Press is a department of the University of Oxford.
It furthers the University's objective of excellence in research, scholarship,and education by
publishing worldwide in

Oxford New York

Auckland Cape Town Dar es Salaam Hong Kong Karachi
Kuala Lumpur Madrid Melbourne Mexico City Nairobi
New Delhi Shanghai Taipei Toronto

With offices in

Argentina Austria Brazil Chile Czech Republic France Greece
Guatemala Hungary Italy Japan Poland Portugal Singapore
South Korea Switzerland Thailand Turkey Ukraine Vietnam

British Library Cataloguing in Publication Data

Data available

ISBN 978 019 912895 2

10 9 8 7 6 5 4 3 2 1

Printed in Great Britain by Bell and Bain Ltd, Glasgow

Cover photo: SugaAngel/Shutterstock

Mixed Sources
Product group from well-managed
forests and other controlled sources
www.fsc.org Cert no. TT-COC-002769
© 1996 Forest Stewardship Council

FSC

1 Work out these multiplications.

    a 2×14 **b** 5×12 **c** 11×9

    d 10×7 **e** 20×8 **f** 25×7

2 Here are some fraction cards.

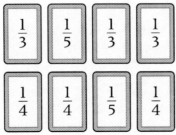

    Use five of the fraction cards to make a total of $1\frac{1}{2}$.

3 Fill in the gaps to show what the units measure.
The first one is done for you.

    a Centimetres measure *length*
    b Kilograms measure _____
    c Litres measure _____
    d Square metres measure _____

4 Use a compass and ruler to draw a triangle that has these
side lengths: 6 cm, 6 cm, 9 cm.

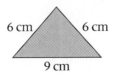

5 Write down events that are impossible, unlikely, evens,
likely and certain.
An example of an impossible event is:
It is impossible for humans to breathe underwater.

1 a Write each of these numbers in figures.

   i One thousand and twenty
   ii Thirty thousand and two hundred
   iii Forty-four thousand and forty-four
   iv Three million and ten

b Write each of these numbers in words.

   i 2300 **ii** 34 500 **iii** 102 000
   iv 255 900 **v** 999 000 **vi** 2 345 000

2 Calculate each of these.

   a 14×10 **b** $4500 \div 100$ **c** 2.6×100
   d 0.34×1000 **e** $156.7 \div 10$ **f** 4.56×10
   g $356 \div 10$ **h** $56\,000 \div 100$ **i** 0.54×1000

3 Write the number each of the arrows is pointing to.

a

```
40    45    50    55    60
```

b

```
40    50    60    70    80
```

c

```
400       450       500
```

d

```
500       600       700
```

4 This is the Fowey–St Austell bus timetable.

Fowey	11.10	11.40
Tywardreath	11.22	11.52
Par	11.29	11.59
St Blazey	11.36	12.06
St Austell	11.58	12.28

a What time does the 11.10 bus from Fowey arrive in St Austell?
b A bus leaves Par at 11.59. How long does it take to get to St Austell?

1 Write these temperatures in order, starting with the coldest temperature. All temperatures are in degrees Celsius.

a	−4	9	0	−6	12	−15
b	−1	1	5	−9	3	10
c	12	3	−2	3	0	−4
d	−5	10	−35	13	3	20
e	0	−10	9	3	−5	−20
f	−100	30	0	−29	39	12

Example

The daytime temperature in Mansfield was 4°C.
Overnight the temperature fell by 8°C.
What was the night-time temperature in Mansfield?

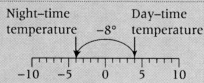

Night–time   Day–time
temperature   −8°   temperature

Use the number line to countdown 8° from the 4° start.
The night-time temperature in Mansfield was −4°C.

2 What temperature is

  a 4 degrees lower than 6°C
  b 10 degrees higher than −1°C
  c 8 degrees lower than 0°C
  d 4 degrees higher than −5°C
  e 14 degrees lower than −3°C
  f 12 degrees higher than −19°C

3 Work out

  a 4 − 5   **b** −3 + 4   **c** −4 + −5   **d** 15 − 18
  e 14 − −4   **f** −4 − −3   **g** −3 + 15   **h** −10 − 3
  i 14 + −14   **j** −8 − −3   **k** −7 + −4   **l** 6 + −6

4 Work out these multiplications.

  a −4 × 5   **b** 9 × −3   **c** −11 × 0   **d** −5 × −4
  e 10 × −5   **f** −4 × −4   **g** 5 × −3   **h** −6 × −6

1 Round each of these numbers to the nearest ten.

   a 56   **b** 67   **c** 184   **d** 365
   e 436   **f** 4192   **g** 97   **h** 32.13

2 Round each of these numbers to the nearest hundred.

   a 453   **b** 192   **c** 76   **d** 983
   e 3458   **f** 934.54 **g** 9875   **h** 17 494

3 Round each of these numbers to the nearest thousand.

   a 4556   **b** 9432   **c** 35 467   **d** 87 567
   e 294 578 **f** 99 483 **g** 99 786 **h** 675 895

4 Use rounding to estimate the answer to each of these. Show how you rounded the numbers.

   a 56 + 182   **b** 598 − 86   **c** 1348 + 67
   d 1504 − 497 **e** 4355 + 897

5 Lawson wrote the temperatures at different times on 1st February 2006.

Time	Temperature in °C
Midnight	−7
5 am	−10
11 am	1
4 pm	6
9 pm	−2

   a Write

      i the highest temperature
      ii the lowest temperature

   b Work out the difference in temperature between
      i midnight and 5 am
      ii 5 am and 4 pm.

At 11 pm the temperature has fallen by 4°C from its value at 9 pm.

   c Work out the temperature at 11 pm.

6 Write the number that each arrow is pointing to.

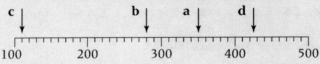

1 Write each of these numbers in figures.
  a Two thousand and five
  b Forty-two thousand and five hundred
  c Ninety-nine thousand nine hundred
  d Seven million and seventy-seven

2 Write each of these numbers in words.

a 5700	**b** 14 800	**c** 455 000
d 200 005	**e** 800 800	**f** 8 648 000

Example

Calculate

  a 35×10　　**b** 2.4×100　　**c** $36.8 \div 10$

  a $35 \times 10 = 350$　　　　　Hint: Use a place value
  b $24 \times 100 = 240$　　　　　　　table to help you.
  c $36.8 \div 10 = 3.68$

3 Work out

a 45×100	**b** $3600 \div 10$	**c** 6.8×10
d 0.77×100	**e** $678.5 \div 10$	**f** 9.96×100

4 Work out

a $6 - 8$	**b** $-7 + 8$	**c** $-8 + -9$
d $9 - 16$	**e** -9×5	**f** 6×-6
g -1×10	**h** -7×-9	

5 Write these temperatures in order starting with the coldest temperature. All temperatures are in degrees Celsius.

a	-6	11	3	-4	12	-16
b	-7	8	0	-7	3	1
c	11	-6	9	-5	2	-14
d	-15	6	-16	5	13	2
e	13	-11	19	-6	-4	-2
f	-10	-3	10	-9	15	14

1 Use the words impossible , unlikely , even chance ,
   likely or certain to describe these outcomes.

   a You will get an even number when you roll
      an ordinary dice.
   b You will be able to fly.
   c It will snow in winter.
   d It will rain during July.
   e You will have a birthday this year.
   f There are 32 days in December.
   g You will pick a white ball from a bag containing
      8 black and 8 white balls.
   h You will become famous.
   i You will spin an even number on a spinner
      with the numbers 2,3,4,5,6.
   j You will pick a milk chocolate sweet from a
      bag containing 5 milk chocolates and 7 plain chocolates.

2 Put these events in order of the chance of landing on white,
   starting with impossible and finishing with certain.

   a   b   c   d   e

3 A bag of sweets contains these flavours

   4 strawberry   3 orange   2 blackcurrant   5 lemon

   Calculate the probability of picking

   a an orange sweet   **b** a lemon sweet
   c a blackcurrant sweet **d** a raspberry sweet.

4 The letters of the word HELLO are put into a bag. One letter
   is taken out at random.
   Calculate the probability that the letter is

   a an H   **b** an L   **c** a vowel
   d a consonant   **e** an O   **f** a B.

1 An ordinary dice is rolled. What is the probability that the dice will land on

   a a 6   **b** an odd number
   c a square number   **d** an 8
   e a prime number   **f** a multiple of 3?

2 A spinner is numbered 2, 2, 2, 3, 4, 5.

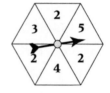

   a Which number is the spinner most likely to land on?
   b What is the probability that the spinner will land on

     i a 2
     ii a number greater than 3
     iii an even number
     iv a 6?

3 The letters of the word MATHEMATICS are put into a bag. One letter is taken out at random.
What is the probability that the letter is

   a an M   **b** an A   **c** a vowel
   d a consonant   **e** a T   **f** a B?

4 A bag contains 3 grey balls, 4 white balls and 5 black balls. One ball is taken out of the bag without looking. What is the probability that the ball is

   a white
   b black
   c grey
   d blue?

1 A packet contains only yellow counters and green counters.
There are 8 yellow counters and 4 green counters.
A counter is to be taken from the packet at random.
Write the probability that

   a a yellow counter will be taken
   b a yellow counter will not be taken.

2 An ordinary dice is to be thrown once.

   a Write down the probability of the dice landing on

     i a four   **ii** an odd number.
  A second dice is to be thrown once.
  The probability that this dice will land on each of the
  numbers 1 to 6 is given in the table.

Number	1	2	3	4	5	6
Probability	x	0.2	0.3	0.1	0.2	0.1

   b Find the probability of the dice landing on a 1.
   c Find the probability that the dice will land on a number
    higher than 4.

3 The two-way table shows the number of students who study
either French or Spanish.

	French	Spanish
Boy	8	12
Girl	13	9

   a How many students study Spanish?
   b A student is selected at random. What is the probability
    they will be

     i a boy
     ii a girl who studies French
     iii a boy who studies Spanish
     iv a student who studies French?

1 A bag contains 5 grey balls, 2 white balls and 3 black balls.
One ball is taken out of the bag without looking.

What is the probability that the ball is

a white   **b** black   **c** grey   **d** red?

2 These cards are shuffled and a card is picked at random.

Find the probability that a card picked at random will be

a a 10 of Hearts   **b** a 10
c a Heart   **d** a King.

3 Describe an event that is

a certain   **b** unlikely   **c** even chance
d impossible   **e** likely.

4 The two-way table shows the number of students who study either Geography or History.

	Geography	History
Boy	14	16
Girl	13	18

a How many students study History?
b A student is selected at random. What is the probability they will be

   i a boy
   ii a girl who studies Geography
   iii a boy who studies History
   iv a student who studies Geography?

1 a Write these numbers in figures.

    i three thousand and five
    ii twenty-two thousand, four hundred and four
    iii eleven thousand and twenty-two
    iv one million and one hundred
    v nine hundred and ninety-nine thousand and
      ninety-nine

b Write these numbers in words.

    i 23 000　　**ii** 20 004
    iii 345 900　**iv** 2 000 000
    v 3 500 005　**vi** 399 999

2 a Round each number to the nearest ten.

    i 46　　**ii** 89
    iii 134　**iv** 198
    v 55　　**vi** 468

b Round each number to the nearest hundred.

    i 246　　**ii** 198
    iii 245　**iv** 589
    v 4698　**vi** 9988

c Round each number to the nearest whole number.

    i 9.7　　**ii** 1.2
    iii 18.2　**iv** 23.8
    v 2.55　**vi** 10.09

3 Calculate each of these.

    a 65×100　　**b** $4200 \div 100$　　**c** 7.9×100
    d 0.68×10　**e** $678.5 \div 100$　**f** 9.96×10

1 Ryan did a survey to find out the number of brothers and sisters the students in his class have. The results were

3	4	0	3	1	3	0	0	2
2	5	4	3	3	2	4	2	3
2	0	5	2	2	3	0	2	4

a Copy and complete the tally chart to show this information.

Number of brothers and sisters	Tally	Number of students
0		
1		
2		
3		
4		
5		

b How many students have more than three brothers and sisters?
c Work out how many brothers and sisters the class have in total.

2 A spinner is spun 18 times to obtain these results.

3	5	5	4	2	4	3	3	5
3	3	4	5	2	5	4	5	3

a Draw and complete a tally chart to show these data.
b Calculate the total score from all 18 spins.

3 The number of GCSEs that students obtained are shown.

5	6	4	7	4	6	7	8	9
4	6	7	8	5	5	6	5	7
6	5	4	7	6	5	5	4	9

a Construct a suitable data-collection sheet to show these data.
b What was the most common number of GCSEs obtained?
c Calculate the total number of GCSEs obtained.

1 Andrew is a fisherman and wants to find out which fish people prefer. He asks 30 people to choose their favourite from cod (C), mackerel (M), plaice (P) and sole (S). The results were:

C	C	M	P	P	S	C	C	P	P
C	M	C	P	M	C	C	P	C	C
C	C	M	P	M	C	P	P	S	S

a Copy and complete this frequency table to show the results.

Type of fish	Frequency
Cod	

b State the favourite type of fish.

2 Andrew uses a questionnaire to find out what people think about eating fish.

a One question in his questionnaire is:
   How old are you?
   Give one suggestion of how this question could be improved.

b Another question in his questionnaire is:
   Don't you think that eating fish is good for you?
   Give one criticism of this question.

3 The speeds of tennis serves, in mph, in a tennis match are shown.

98	112	86	89	95	103	109	99	107
95	89	91	102	115	108	90	114	89
105	100	98	88	114	109	93	91	101

a Copy and complete this frequency table for these data.

Speed of serve (mph)	Number of serves
85–89	
90–94	
95–99	

b Serves over 100 mph are very difficult to return. Find the number of serves over 100 mph.

c Which class interval has the most common speed?

1 Year 7 students during autumn term study either History, Geography or RE.
The two-way table shows information about these students.

	History	Geography	RE	Total
Male	12		18	
Female	21		17	46
Total		32		

a Complete the two-way table.
b How many students were in Year 7?
c How many were female?
d How many students studied Geography or History?

2 The two-way table shows the number of students in a class who wear glasses.

	Wear glasses	Don't wear glasses
Boy	6	11
Girl	4	9

a State the number of students who

   i do not wear glasses and are female
   ii wear glasses and are male.

b Find the number of students who

   i are female   **ii** wear glasses.

c Find as a simplified fraction, the proportion of students who

   i wear glasses   **ii** are girls that wear glasses.

3 Explain the difference between primary and secondary data.

1 The two-way table shows the number of students in a class who are left and right handed.

	Left-handed	Right-handed
Male	4	14
Female	6	16

a State the number of students who are
   i female and left-handed   ii right-handed males.

b Find the number of students who are
   i female   ii left-handed.

c Giving your answer as a fraction in its simplest form, find the proportion of students who are
   i right-handed   ii left-handed boys.

2 a Write the number 3056 in words.
   b Write what the 5 stands for in 3056.

3 Use rounding to estimate the answer to each of these, showing how you rounded the numbers.

   a $96 + 171$　　　　b $459 - 58$
   c 49×23　　　　d $149 \div 32$
   e $342 - 129$　　　　f 32×86

4 Work out:

   a $3 - 5$　　　　b $-2 + 3$　　　　c $-4 + 7$
   d $7 - 9$　　　　e $4 - -2$　　　　f $5 - -1$
   g $-4 - 5$　　　　h $-2 - 3$　　　　i $-3 - -6$

1 Using a written method work out these calculations.

a 345 + 134	**b** 734 + 154	**c** 456 + 345
d 538 + 122	**e** 672 + 245	**f** 938 + 267
g 498 + 356	**h** 932 + 834	**i** 1234 + 546
j 5643 + 347	**k** 8567 + 456	**l** 9875 + 1896

2 Using a written method work out these calculations.

a 456 − 234	**b** 857 − 345	**c** 984 − 232
d 867 − 123	**e** 345 − 126	**f** 567 − 329
g 459 − 282	**h** 1893 − 924	**i** 345 − 286
j 3458 − 535	**k** 8947 − 5958	**l** 4587 − 3589

3 Use the digits 4, 5, 6, 7 and 9 to complete each addition problem.

a
$$\begin{array}{r} ??? \\ + \ ?? \\ \hline 724 \end{array}$$

b
$$\begin{array}{r} ??? \\ + \ ?? \\ \hline 1021 \end{array}$$

c
$$\begin{array}{r} ??? \\ + \ ?? \\ \hline 553 \end{array}$$

4 This table shows the numbers of cars, vans and lorries on the Isle of wight ferry one week.

	Cars	Vans	Lorries
Monday	92	68	24
Tuesday	57	95	32
Wednesday	103	87	41
Thursday	86	67	37
Friday	49	93	52
Saturday	127	41	6
Sunday	141	52	18

  a How many lorries used the ferry in the whole week?
  b How many vehicles used the ferry on Thursday?
  c How many vans used the ferry at the weekend?
  d How many cars used the ferry on Monday to Friday?
  e Which day did the ferry carry the most vehicles?

1 Use an appropriate mental method to match the questions with the answers. What do the letters representing the answers spell?

55×8, 5×25, $88 \div 2$, 6×16, 5×32,
$96 \div 3$, 3×124, 12×10, $105 \div 5$

S = 96	E = 120	O = 44	L = 372	S = 125
E = 32	I = 440	S = 21	C = 160	

2 Use an appropriate mental method to solve each problem.

a Kai buys 5 kg of fish at £1.50 a kg.
How much does it cost?

b Pam shares £65 between her 5 grandchildren.
How much does each child receive?

c Steve buys 8 bars of chocolate. Each bar cost 24p.
How much money does he spend?

d James runs 400 metres in 80 seconds.
How long does it take him to run 10 metres?

e Pritesh hits 12 golf balls each 200 metres.
How far has he hit the 12 balls in total?

f Alex walks her dog 3000 metres in 20 minutes.
How far does she walk in 1 minute?

3 Work out each of these.

a $3 + 4 \times 5$ **b** $5 + 2 \times 6 + 3$ **c** $4 + 2 \times (3 + 5)$
d $4^2 - 16 \div 2$ **e** $18 \div 3 - 2 \times 2$ **f** $(5 + 10) \div 3$
g $\dfrac{6 + 4 \times 3}{3}$ **h** $\dfrac{4 \times 5 + 2}{11}$ **i** $\dfrac{27}{2^2 + 5}$

4 Copy each of these calculations and insert brackets, where necessary, to make each statement correct. Not every statement may need brackets.

a $2 + 3 \times 4 = 14$ **b** $6 \div 4 - 2 = 3$
c $7 \times 2 + 8 \div 4 = 16$ **d** $3^2 - 4 \times 2 + 6 = 40$
e $40 \times 2^2 - 1 = 120$ **f** $3 + 2 \times 9 = 21$
g $5 + 20 \div 4 = 10$ **h** $4^2 + 3 \times 2 = 38$

1 David takes 14 boxes out of his van.
The weight of each box is 26 kg.

a Work out the total weight of the 14 boxes.

He then fills the van with wooden crates.
The weight of each crate is 56 kg.
The greatest weight the van can hold is 800 kg.

b Work out the greatest number of crates that the van can hold.

2 Use your calculator to work out

$$(2.4^2 + 1.9) \times 2.38$$

a Write all the figures shown on your calculator display.
b Round the answer in part **a** to 2 decimal places.

3 Teresa bought 48 cans of drink for 45p each.

a Work out the total amount she paid.

Teresa sold all the drinks for a total £26.40.
She sold each can for the same price.

b Work out the price at which Teresa sold each can.

4 A bar of chocolate costs 24p.

a Write the greatest number of bars of chocolate that you can buy for £3 (300p).

Lorraine buys 12 bars of chocolate.
She pays with a £5 note (500p).

b How much change should she get?

Rachael buys some bars of chocolate.
She pays with a £10 note. She receives £1.36 change.

c How many bars of chocolate does she buy?

1 Work out these using a written method.

    a $124 + 456$ **b** $586 - 145$ **c** $156 + 998$

    d $345 - 278$ **e** 45×14 **f** $180 \div 15$

    g 34×12 **h** $294 \div 14$ **i** 345×16

2 a A packet of crisps costs 28p.

    i Find the greatest number of packets of crisps that you can buy for £5.

    ii How much change will you receive from the £5?

  b Michelle buys 14 cans of drink at 45p each for her football team. She pays with a £10 note.

    i How much change should she get?

    ii List the coins she would receive if the change was given in the smallest number of coins.

3 A dice is to be thrown.
The probability that this dice will land on each of the numbers 1 to 6 is given in the table.

Number	1	2	3	4	5	6
Probability	0.3	0.1	0.1	0.2	0.1	x

The dice is to be thrown once.

  a What is the probability of it landing on a 6?

  b What is the probability that the dice will land on a number less than 5?

4 Use a written method to work out these.

    a 17×13 **b** 29×41 **c** 14×236

    d $652 \div 4$ **e** $672 \div 7$ **f** $364 \div 13$

    g 6.5×3 **h** 32×2.5 **i** $113.75 \div 7$

1 The pictogram shows the number of different colour cars that passed a school during one day.

Red	☐ ☐ ☐ ☐
Silver	☐ ☐ ☐ ☐ ☐
Blue	☐ ☐
Black	
Other	

Key: ☐ represents 4 cars

a Copy and complete the pictogram to show 8 black cars and 14 other cars.

b Which colour car was the most popular and how many of this colour car were seen?

c How many cars were seen in total?

2 The number of goals scored by five players in a hockey team are shown in the frequency table. Draw a bar chart to show this information.

Player	Number of goals
Heidi	4
Leanne	8
Claire	6
Sonia	6
Sue	13

3 The favourite types of films for boys and girls are shown in the bar chart.

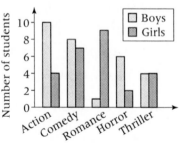

Type of film

a Action was the favourite type of film for how many

   i boys   **ii** girls?

b What type of film did more girls like than boys?

c Calculate the total number of boys and girls in the survey.

1 36 people took part in a survey. They were asked which type of holiday they prefer out of these categories: camping, beach, winter, adventure or sport. The results are shown in the table.

Type of holiday	Number of people
Camping	5
Beach	14
Winter	7
Adventure	4
Sport	6

Draw a pie chart to show this information.

2 James received £90 for his birthday and he decided to spend it as shown in the table. Draw a pie chart to show this information.

Computer game	£36
Sweets	£2
Trainers	£29
DVD	£17
T-shirt	£6

3 A fruit seller sells 180 pieces of fruit in a day, as shown in the pie chart.

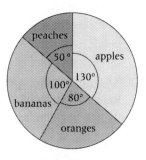

a Calculate the angle that represents one piece of fruit.

b Find the number of

   i bananas   **ii** peaches sold.

c State the modal fruit sold.

4 The pie chart shows the favourite sports for 90 students.
Find

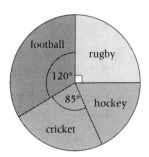

a the angle for rugby

b the angle for hockey

c the angle that represents 1 student

d the number of students who like

   i rugby   **ii** hockey   **iii** cricket   **iv** football.

1 The resting pulse rates, in beats per minute (bpm), of 30 athletes are given.

56	74	63	72	83	49	58	59	79	48
73	77	89	57	64	61	69	75	76	81
70	72	84	52	44	57	74	75	81	77

Copy and complete the stem-and-leaf diagram.

Key: 4 | 9 means 49 bpm

4	
5	
6	
7	
8	

2 The speeds, in mph, of bowling in a cricket match are shown.

| 67 | 87 | 75 | 65 | 89 | 65 | 98 | 56 | 75 | 84 | 88 | 57 |
| 67 | 89 | 76 | 75 | 84 | 85 | 76 | 74 | 83 | 67 | 69 | 72 |

Draw a stem-and-leaf diagram using stems of 50, 60, 70, 80 and 90. Remember to give a key.

3 The number of text messages Sam sent over a week is given.

Day	Mon	Tues	Wed	Thurs	Fri	Sat	Sun
Number of text messages	7	12	8	17	23	38	19

Draw a line graph to show this information.

4 The attendance, to the nearest 1000, at premiership football games in the 90s is shown.

Years	1992–3	1993–4	1994–5	1995–6	1996–7
Attendance	21 000	23 000	24 000	28 000	28 000

a Draw a line graph to show this information.
b Between which two years was there the biggest increase in attendance?

1 The GCSE Maths and English results for six years are represented by these line graphs.

Make two statements about the Maths and English GCSE results over the 6 years.

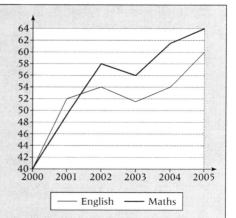

English ——— Maths

2 Here are the times, in minutes, taken to finish a jigsaw.

27	39	42	22	39	41	39	40
28	39	51	19	43	39	42	45
18	20	28	42	35	38	32	39

Draw a stem-and-leaf diagram to show these times.

3 Charlotte did an investigation into her friends' favourite sports and recorded the results in this chart.

Write down two things that are wrong with Charlotte's bar chart.

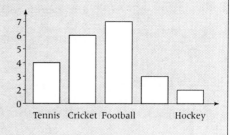

4 The numbers of goals scored over the season by five players in a football team are shown in the frequency table.

Draw a vertical bar chart to show this information.

Player	Number of goals
Wayne	4
David	12
Frank	9
Michael	16
Stephen	15

1 State one event that is

    a impossible   **b** unlikely   **c** evens
    d likely   **e** certain.

2 The letters of the word ISOSCELES are put into a bag.
A letter is taken out at random.
What is the probability of picking

    a an E   **b** an S   **c** a vowel?

3 The exam performance of a school over 6 years is shown.

Year	2000	2001	2002	2003	2004	2005
% A–C	34%	38%	35%	40%	45%	46%

    a Draw a line graph, on a copy of the grid, to show this information.

    b Describe what has happened to the %A–C over the 6 years.

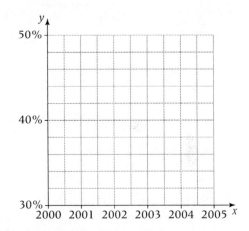

4 Copy and complete this addition square.

+	36	27	87	?	92
15				46	
?			106		
45					
?	49				
56					

1 Calculate the mean for each set of numbers.

  a 4, 5, 5, 6
  b 12, 9, 10, 8, 11
  c 105, 102, 95, 98, 101, 99
  d 6, 9, 8, 13, 0
  e 1, 4, 4, 4, 4, 5, 6
  f 0, 0, 3, 2, 5, 12, 4, 4, 6
  g 3, 4, 5, 5, 4, 2, 6, 7

2 At Rose Cottage Bed and Breakfast, the nightly rate changes each month, depending on the season.
The table gives the nightly rate during each month of the year.

Month	Jan	Feb	Mar	Apr	May	Jun	Jul	Aug	Sep	Oct	Nov	Dec
Price in £	35	35	42	45	50	55	60	65	50	40	35	40

Assuming the same number of nights are available per month, what is the Bed and Breakfast's mean nightly rate.

3 Work out the median for each set of numbers.

  a 7, 8, 12, 4, 3
  b 1, 5, 8, 7, 4, 1, 7, 4, 3
  c 0.45, 0.38, 0.12, 0.50, 0.75
  d 32, 35, 33, 26, 37, 54, 23, 26
  e 5.4, 4.6, 5.1, 4.5, 4.9, 5.3
  f £3.45, £8.45, £0.45, £9.03, £4.59, £3.99
  g $1\frac{1}{2}$, 3, $4\frac{1}{2}$, 5, $3\frac{1}{4}$, 4, 5, 3, $4\frac{3}{4}$, 5

4 Ten students took a test and their results were

  45% 67% 94% 83% 29% 68% 72% 89% 35% 59%

  a Calculate the mean score.
  b Work out the median score.

1 Calculate the range and mode of each of these sets of numbers.

  a 7, 8, 12, 8, 3
  b 1, 5, 8, 4, 4, 1, 7, 4, 3
  c 0.45, 0.75, 0.12, 0.50, 0.75
  d 32, 35, 33, 26, 37, 54, 23, 26
  e 5.4, 4.5, 5.1, 4.5, 4.9, 5.3
  f £3.45, £0.45, £0.45, £9.03, £4.59, £3.99
  g $1\frac{1}{2}$, 3, $4\frac{1}{2}$, 5, $3\frac{1}{4}$, 4, 5, 3, $4\frac{3}{4}$, 5

2 Julie added up the number of pages she read each month last year and recorded the data in this table.

Month	Jan	Feb	Mar	Apr	May	Jun	Jul	Aug	Sep	Oct	Nov	Dec
Number of pages read	35	35	42	45	50	55	60	65	50	40	35	40

  Find the **a** range **b** modal number of pages read.

3 The number of goals scored by a team over 10 games is given.

  4, 3, 2, 1, 1, 4, 0, 1, 1, 3

  a Draw and complete a frequency table using these headers.

Number of goals	Tally	Frequency

  b Calculate the mean, median, mode and range for the teams.

4 Ten people were asked how many films they had watched in the last week. The results are shown in the tally chart.

Number of films	Tally	Frequency
0	I	1
1	JHT I	6
2	II	2
3	I	1

  Calculate the mean, median, mode and range of the ten numbers.

  Hint: It may make it easier to write the ten numbers in a list.

1 The number of goals scored by two hockey teams, in ten games, is show in the two bar charts.

Lutterworth hockey team

Rugby hockey team

a List the number of goals scored by each of the two teams.
b Calculate the mean, median, mode and range for the two teams.
c Compare the two sets of results. Which team do you think is the better team? Explain your answer.

2 The stem-and-leaf diagram shows pulse rates after exercise. Calculate the **i** mean **ii** median **iii** mode **iv** range.

6	7
7	3 7 9
8	2 4 8 8
9	2 2 3 4 5 7 9 9
10	2 4 6 7
11	0 1

Key: 7|3 means 73 beats per minute

3 The ages, in years, of 20 teachers are recorded

27 45 36 23 52 48 39 40 32 55
29 37 56 25 42 41 36 44 36 51

a Copy and complete an ordered stem-and-leaf diagram.

2	
3	
4	
5	

Key: 2|7 means 27 years

b Calculate the **i** mean **ii** mode **iii** median **iv** range.

1 Find the mean, median, mode and range for these sets of data.

   a 2, 3, 4, 3, 3
   b 12, 15, 8, 5, 10, 10
   c 6, 7, 8, 2, 2, 5, 6, 4, 2, 8
   d 14, 17, 14, 13, 18, 14

2 Thirty people were asked how many bars of chocolate they had eaten in the last week. The results are shown in the tally chart.

Number of bars of chocolate	Tally	Frequency
0	ЖШ І	
1	ЖШ ЖШ ІІ	
2	ЖШ	
3	ІІ	
4	ІІ	
5	ІІІ	

   a Copy and complete the frequency table.
   b Calculate the mean, median, mode and range.

3 Mrs Edwards gives her class a maths test.
Here are the test marks for the girls.

7, 5, 8, 5, 2, 8, 7, 4, 7, 10, 3, 7, 4, 3, 6

   a Work out the median.   **b** Work out the range.

The median mark for the boys was 7 and the range of the marks of the boys was 4.

   c By comparing the results explain whether the boys or girls did better in the test.

4 Five numbers have a mean of 10, median of 10, mode of 10 but a range of 4.

   a Give five possible numbers.
   b Find more sets of possible numbers.

1 The time series graph shows gas bills over 7 quarters.

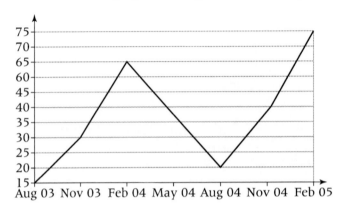

Aug 03 Nov 03 Feb 04 May 04 Aug 04 Nov 04 Feb 05

a What do you notice about the bills in August compared to February?
b Which bill is the **i** lowest **ii** highest?
c Between which two quarters is there the greatest increase in gas bills?

2 The table shows the daytime and nightime temperatures recorded in Manchester one week.

	Day temp (°C)	Night temp (°C)
Monday	2	−4
Tuesday	3	−2
Wednesday	1	−5
Thursday	0	−5
Friday	1	−3

a What time of year do you think these temperatures were recorded?
b Which night had the highest temperature?
c Which day had the lowest temperature?
d How much did the temperature fall between daytime Wednesday and nighttime Wednesday?
e Which day had the greatest difference between day and night temperatures?

1 Copy this rectangle 5 times. Shade each rectangle to show these fractions.

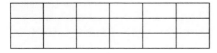

  a $\frac{1}{2}$ **b** $\frac{3}{4}$ **c** $\frac{1}{3}$ **d** $\frac{4}{9}$ **e** $\frac{5}{6}$

2 a There are 34 people on a bus. 16 are male and the rest are female. What fraction of the people on the bus are
  i male   **ii** female?

  b Manjit earns £450 a week. She pays £110 of her money in tax. She saves £90 each week. The rest she spends. What fraction of her weekly wage does Manjit
  i pay in tax   **ii** save   **iii** spend?

  c Rhiannon has 13 pairs of trousers, 2 skirts, 25 tops and 9 pairs of shoes. What fraction of her clothes are
  i trousers   **ii** tops   **iii** skirts?

3 Write these percentages as fractions in their simplest form.

  a 75%   **b** 10%   **c** 25%   **d** 30%
  e 8%   **f** 60%   **g** 45%   **h** 20%

4 Write each of these fractions as a percentage.

  a $\frac{45}{100}$ **b** $\frac{13}{50}$ **c** $\frac{4}{25}$
  d $\frac{7}{10}$ **e** $\frac{3}{5}$ **f** $\frac{3}{20}$
  g $\frac{46}{200}$ **h** $\frac{85}{500}$ **i** $\frac{45}{150}$

5 In a packet of 25 biscuits, 12 are milk chocolate and the rest are plain chocolate. What percentage are plain chocolate?

1 Write these percentages as decimals

   a 42%   **b** 80%   **c** 90%   **d** 3%

2 Write these decimals as fractions in their simplest form.

   a 0.2   **b** 0.8   **c** 0.45

   d 0.35   **e** 0.09   **f** 0.125

3 Change these fractions to decimals without using a calculator.

   a $\frac{6}{10}$ **b** $\frac{1}{4}$ **c** $\frac{10}{25}$

   d $\frac{34}{50}$ **e** $\frac{9}{10}$ **f** $\frac{22}{200}$

4 Change these fractions to decimals using a calculator.
Give your answers to a suitable degree of accuracy.

   a $\frac{13}{40}$ **b** $\frac{5}{8}$ **c** $\frac{12}{15}$

   d $\frac{3}{7}$ **e** $\frac{19}{21}$ **f** $\frac{12}{7}$

5 Use a suitable method to calculate these.

   a $\frac{3}{10}$ of £30   **b** $\frac{5}{7}$ of 35 kg   **c** $\frac{4}{9}$ of 63p

   d $\frac{7}{20}$ of 100 dogs   **e** $\frac{7}{15}$ of 45°   **f** $\frac{12}{25}$ of €400

   g $\frac{4}{5}$ of £300   **h** $\frac{9}{13}$ of 104 g   **i** $\frac{5}{6}$ of 192 people

6 Sharanjit does a survey of 150 people to find out their
favourite colour. She works out the fraction of people who
like each colour. Copy and complete the table working out
the frequencies for each colour.

Colour	Frequency	Fraction of total
Red		$\frac{1}{5}$
Blue		$\frac{9}{30}$
Purple		$\frac{1}{6}$
Green		$\frac{1}{15}$
Other		$\frac{4}{15}$

Example

Calculate these percentages without using a calculator.

a 25% of €260 **b** 20% of 330 kg **c** 30% of £240

a 25% of €260

$= \frac{1}{4}$ of 260

$= 260 \div 4 = 65$

$= €65$

b 20% of 330 kg

$= \frac{1}{5}$ of 330

$= 330 \div 5 = 66$

$= 66$ kg

c 30% of 240

10% of 240 = 24

So $3 \times 24 = 72$

$= £72$

1 Find these percentages without using a calculator.
You must show all of your workings.

    a 50% of £300   **b** 25% of £300   **c** 10% of £40

    d 1% of 500p   **e** 10% of €450   **f** 20% of $250

    g 40% of 70p   **h** 15% of $900   **i** 55% of 1500 g

    j 75% of 660 kg   **k** 35% of €150   **l** 17.5% of 120p

2 a Claire used 2000 letters in her essay. 35% of the letters were vowels. How many of the letters were vowels?

   b Amir earns £35 a week. He spends $\frac{7}{10}$ of what he earns. How much does he spend?

   c A tank can hold 140 litres when full. Water is poured into the tank until it is 65% full. How much water is in the tank?

3 Find these percentage increases and decreases.

   a Increase £30 by 10%   **b** Decrease 150 m by 20%

   c Decrease £66 by 5%   **d** Increase 720 kg by 25%

   e Increase £36 by 1%   **f** Decrease 450p by 15%

4 Find these correct to 2 decimal places.

   a Increase £56 by 12%   **b** Decrease $49 by 22%

   c Decrease 39 kg by 36%   **d** Increase 230 g by 38%

   e Increase €560 by 7%   **f** Decrease 1340 m by 73%

   g A computer costs £499 before VAT is added. VAT is 17.5%. Increase the cost of the computer by 17.5% to find the cost of the computer when the VAT is added.

   h A holiday costs £2695 but it is reduced by 9% in the sale. Work out the cost of the holiday when it is in the sale.

1 Put these in order of size starting with the smallest.

a $\frac{3}{4}$, 70%, $\frac{7}{8}$, 0.72　　　**b** 0.4, 45%, $\frac{4}{9}$, $\frac{3}{10}$

c 34%, $\frac{2}{5}$, 25%, 0.3　　　**d** 98%, 1, $\frac{9}{10}$, 0.99

2 a Michelle took three tests. In maths she scored 59 out of 72, in Science she scored 45 out of 60 and in RE she scored 27 out of 50.

i In which subject did she do best?

ii In which subject did she do worst?

b Leah took three different tests. In English she scored 52 out of 80, in Graphics she scored 54 out of 70 and in PE she scored 52 out of 60.

i In which subject did she do best?

ii In which subject did she do worst?

3 Write each of these ratios in their simplest form.

a 2 : 8　　　**b** 5 : 55　　　**c** 7 : 70

d 10 : 100　　**e** 12 : 16　　**f** 18 : 45

g 72 : 9　　　**h** 48 : 12　　**i** 25 : 125

4 a Forid buys 5 adult tickets for the theatre. He pays £62.50. Work out the cost of

i 10 tickets　　**ii** 15 tickets

iii 1 ticket　　**iv** 4 tickets.

b Arshi works for 9 hours and gets paid £42.75. She is paid the same amount each hour. How much will she get paid for working

i 18 hours　**ii** 27 hours　**iii** 1 hour　**iv** 35 hours?

c Amy knows there are 5 miles in 8 km. How many miles are there in

i 24 km　　**ii** 64 km　　**iii** 12 km　　**iv** 2 km?

1 a Shade 0.3 of this shape.   **b** Shade 75% of this shape.

 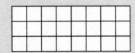

2 What proportion of each of the shapes in Question **1** is shaded? Write each of your answers as a fraction in its simplest form.

3 Write these numbers in order of size.
Start with the smallest number.

   a $\frac{1}{2}, \frac{3}{4}, \frac{4}{5}, \frac{2}{3}, \frac{1}{6}$ **b** $0.43, \frac{2}{5}, 45\%, \frac{1}{2}, \frac{4}{7}$

4 There are 700 students at Fowey School.

   a 15% of these students were absent from school on Friday. Work out how many students were not absent on Friday.

   b Carrie-Ann says that more than 255 of the 700 students were absent on Friday.
Is Carrie-Ann correct? Explain your answer.

   c 56% of the students are girls.
Work out how many of the students are girls.

   d 154 of the students are in Year 7.
Write 154 out of 700 as a percentage.

5 David wants to buy some jeans.
The same jeans are sold in three different shops.

Jeans R Us	Denim Design	Jeans Shop
$\frac{1}{3}$ off usual price of £39	30% off usual price of £36	£20 plus VAT at 17.5%

   a Find the cost of each pair of jeans in the three different shops.

   b Work out the difference in price between the maximum and minimum prices David could pay for the jeans.

1 Calculate these giving your answers correct to 2 decimal places.

  a Increase £67 by 13%

  b Decrease $57 by 62%

  c Decrease 689 kg by 76%

  d Increase 267g by 38%

  e Increase $570 by 8%

  f Decrease 1780 m by 53%

  g A TV costs £299 before VAT is added. VAT is 17.5%. Calculate the cost of the TV after VAT is added.

  h A holiday costs £3125 but it is reduced by 10% in the sale. Work out the cost of the holiday when it is in the sale.

2 The following cards are shuffled and a card is picked at random.

Calculate the probability that a card picked at random will be

  a an ace of spades   **b** a 4

  c a spade   **d** a queen

  e a 7 of diamonds.

3 The number of books 40 students read in one month is recorded in a pictogram.

  a What does represent?

  b What is the modal number of books read by the 40 students?

  c What fraction of students read two or more books a month?

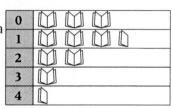

D5 HW2 Probability revision

1 A spinner with the numbers 2, 2, 3, 4, 5, 5, 6, 7 is spun.
What is the probability that the spinner will land on

a an even number   **b** a prime number

c a multiple of 3   **d** a factor of 30

e a square number   **f** not a square number?

2 A pack of 52 playing cards contains four 'suits' which consist
of 13 Spades, 13 Clubs, 13 Hearts and 13 Diamonds.
Each suit has cards numbered 2, 3, 4, 5, 6, 7, 8, 9, 10,
Jack, Queen, King, Ace.
If a card is chosen at random from the pack, what is the
probability it will be

a a Heart

b a Queen

c a Queen of Diamonds

d a 2

e an odd number Spade

f a Jack, Queen or King?

3 Here are two boxes of coloured balls.

Box A   Box B

What is the probability of picking

a a black ball from box A

b a white ball from box A

c a ball that is not grey from box A

d a grey ball from box B

e a ball that is not grey from box B?

Pete wants to pick a grey ball.

f Which box should be pick from to have the greatest
chance of picking a grey ball? Explain your answer.

D5 HW3 Expected and relative frequency

1 The probability that Lutterworth Hockey Team win each game is $\frac{7}{8}$.

    a Calculate the probability that they will lose a game.

    b If they play 24 games, how many of them do you expect them to lose?

2 A bag contains 4 grey, 10 white and 6 black balls.
One ball is taken out at random and then replaced.

    a What is the probability that the ball is

        i grey   **ii** white   **iii** black?

    b If a ball is taken out and replaced 100 times, how many of the balls would you expect to be

        i grey   **ii** white   **iii** not white?

3 Shane carries out a survey about the words in a book.
He chooses a page at random and counts the number of letters in each of the first 50 words. The results are

3	3	5	5	6	7	1	3	4	7
4	7	2	3	4	4	3	5	1	5
2	6	6	5	4	3	4	3	4	1
6	5	5	4	2	5	1	5	6	4
1	4	1	2	6	5	4	4	3	4

    a Draw a frequency table to show these results.
    b State the modal number of letters.
    c A word is chosen at random. What is the probability that it has 4 letters?
    d The book has 10 000 words. Estimate the number of 4-letter words in the book.

1 A coin and an ordinary dice are thrown together.

   a List the 12 possible outcomes, e.g. a Head and a 1.
   b Calculate the probability of getting
      i a Head and an even number **ii** a Head and not a 6.
   c If the coin and dice are thrown 60 times, work out how many times you expect to get a Tail and a number less than 3 together.

2 A fair dice is rolled and a spinner, numbered 1 – 4, is spun.

   a Complete the list of all 24 possible outcomes.

Dice	Spinner
1	1
1	2
1	3
1	4
2	1 etc.

   b What is the probability of getting a double 6?
   c If the dice and spinner were spun 240 times how many times would you expect to get a total of 10?

3 Claire throws a fair coin. She gets a Tail.
Sonia then throws the same coin.

   a What is the probability that Sonia will get a Tail?
   b Sonia throws the coin 40 times. Explain why she may not get exactly 20 Heads and 20 Tails.

4 Jo has a bag of 9 marbles. 2 of the marbles are black and 7 are white. Jo chooses one marble at random.

   She says, 'There are two possible colours, either white or black, so I have half a chance of choosing a white marble and half a chance of choosing a black marble'.

   a Explain why Jo is wrong.
   b Write the probability of Jo choosing each colour.

1 Work out:

    a 2.7 + 3.6   **b** 4.2 + 9.5
    c 8.3 + 2.9   **d** 5.1 + 6.7

2 Work out:

    a 4 × 100   **b** 7 × 1000
    c 7.2 × 10   **d** 5.3 × 100
    e 87 × 100   **f** 4.7 × 10
    g 3.6 × 1000   **h** 37.2 × 100

3 Work out:

    a 2 + 7 + 3 + 4   **b** 2.5 + 3.2 + 2.5 + 3.2
    c 10 + 5.7 + 10 + 5.7   **d** 3.9 + 4.8 + 5

4 Work out:

    a 6 × 5   **b** 3.5 × 4   **c** 8 × 9
    d 2.6 × 5   **e** 4.9 × 10   **f** 5.7 × 6

5 Sonia goes to the shop. She buys

    2 kg of carrots at 15p per kg
    5 oranges at 27p each
    4 kg of potatoes at 56p per kg.

She pays with a £10 note.
Work out how much change she should get.

1 Match these cards into pairs.

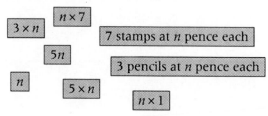

2 There are 9 cakes in a packet.
How many cakes are there in

   a 5 packets   **b** 8 packets
   c x packets   **d** n packets?

> Simplify these expressions.
>
> **a** $p + p + p - p$ **b** $4z - 2z + z$
>
> ··
>
> **a** $p + p + p - p$ **b** $4z - 2z + z$
>
> $= 3p - p$ $= 2z + z$
> $= 2p$ $= 3z$

3 Simplify these expressions.

   a $m + m + m + m$ **b** $a + a + a + a + a + a + a$
   c $b + b + b + b + b$ **d** $p + p + p + p - p - p - p$
   e $r + r + r + r - r - r$ **f** $a + a + a + a + b + b + b$

4 Simplify each of these expressions.

   a $4k - 2k$ **b** $4y - y$
   c $5x + 2x - 3x$ **d** $12j - 5j + 2j$
   e $k - k$ **f** $6p - 5p + 4p + 3p$

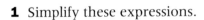

A1 HW3 Collecting terms and substitution

1 Simplify these expressions.

   a $3 \times n$ **b** $n \times n$ **c** $e \div 9$ **d** $4n \div 3$

   e $4 \times a \times b$ **f** $r \times s \times t$ **g** $2 \times a \times b$ **h** $p \div q$

Example

Simplify by collecting like terms.

   a $2x + 6y + 3x + 4y$ **b** $5p + 9q - 5p - 2q$

   a $2x + 6y + 3x + 4y$ **b** $5p + 9q - 5p - 2q$
      $= 5x + 10y$ $= 0p + 7q$
                                $= 7q$

2 Simplify by collecting like terms.

   a $3a + 5b + 2a + 6b$ **b** $8d + 4c + d - c$

   c $10r + 5s + 3r - 2s$ **d** $11p - 3q + 4p - 2q$

   e $9g + 5h - g - h$ **f** $5t + 3s + 6t - 3s$

   g $9w + 4f - 5w - f - 4w$ **h** $11m + 5s - 11m - 5s$

3 Chocolicious makes boxes of chocolates in two sizes.
There are a chocolates in a small box.
There are b chocolates in a large box.
Write an expression for the total number of chocolates in

   a one small box
   b two large boxes
   c two small boxes and three large boxes
   d four small boxes and six large boxes.

4 Work out the value of these expressions.

   a $3a$ when $a = 4$

   b $5d$ when $d = 6$

   c $2a + b$ when $a = 4$ and $b = 3$

   d $3m + n$ when $m = 2$ and $n = 4$

   e $2x - 3y$ when $x = 5$ and $y = 2$

   f $2p + 2q - r$ when $p = 4$, $q = 6$ and $r = 5$

   g y^2 when $y = 3$

   h $x^2 + y$ when $x = 4$ and $y = 2$

1 Calculate the value of each expression when $a = 3$, $b = 4$ and $c = -2$.

a $a + b$ **b** $3a + c$ **c** $a + b + c$

d $3c + a$ **e** $\dfrac{2a + b}{5}$ **f** $\dfrac{2b + 2c}{4}$

g $a \times b$ **h** $a \times b \times c$ **i** $a - b + 2c$

j $\dfrac{3ab}{4}$ **k** $a^2 + 2a + b$ **l** $b^2 + c$

2 The formula $F = \dfrac{9C}{5} + 32$ is used to convert temperatures in °C (Celsius) to °F (Fahrenheit).
Work out the temperature in Fahrenheit (F) when

a $C = 20$
b $C = -10$.

3 Simplify these expressions.

a $x + x$
b $a^2 + a^2 + a^2 + a^2$
c $b \times b$
d $2 \times e \times f$
e $3 \times 5 \times a \times a$

4 Simplify these expressions.

a $b + b + b + b + b$
b $5p + 3p$
c $5t + 3s - 3t + 2s$

1 Write these using algebra.

   a a number multiplied by 6 then 2 added equals 14

   b a number subtracted from 10 equals 3

   c 5 is equal to a number multiplied by 2

   d a number divided by 2 plus 3 equals 8

   e 20 divided by a number is equal to 10

2 Simplify these expressions.

   a $4p + 5q + p + 4q$ **b** $5t - s + 4t + 4s$ **c** $5b + b + 3a - 6b$

   d $k + 4m - 3k + m$ **e** $5g + 3h + h - g$ **f** $5c + 3b + c - 4b + c$

3 Work out the value of each expression.

   a $4d + 4f$ when $d = 5$ and $f = 2$

   b $2m + 6n$ when $m = 2$ and $n = 3$

   c $3p - 5q$ when $p = 10$ and $q = 3$

   d $3e - 2d + f$ when $e = 1$, $d = 2$ and $f = 9$

   e $2b + 3e$ when $b = -4$ and $e = 3$

   f $4t - 5r - s$ when $t = 5$, $r = 2$ and $s = -4$

4 Work out the value of each expression when $a = 4$, $b = 3$ and $c = -5$. Then put the cards in order, smallest value first.

$3a + 4b$ $3b + c$ ab

$2b + 2a + 2c$ $3c + a$ $2c - 2a$

1 Find the missing number in each pair of equivalent fractions.

a $\frac{1}{2} = \frac{?}{12}$ b $\frac{6}{7} = \frac{12}{?}$ c $\frac{5}{9} = \frac{?}{27}$

d $\frac{8}{11} = \frac{24}{?}$ e $\frac{?}{35} = \frac{2}{7}$ f $\frac{6}{?} = \frac{2}{3}$

2 Cancel these fractions into their simplest form.

a $\frac{6}{8}$ b $\frac{3}{12}$ c $\frac{2}{6}$ d $\frac{5}{15}$

e $\frac{4}{12}$ f $\frac{2}{8}$ g $\frac{6}{9}$ h $\frac{4}{10}$

3 Calculate each of these, giving your answer in its simplest form, where appropriate.

a $\frac{1}{2} + \frac{1}{2}$ b $\frac{2}{7} + \frac{4}{7}$ c $\frac{10}{18} - \frac{4}{18}$

d $\frac{4}{15} + \frac{1}{15}$ e $\frac{23}{28} - \frac{2}{28}$ f $\frac{4}{54} + \frac{14}{54}$

4 Write these decimals as fractions in their simplest form.

a 0.2   b 0.7   c 0.45

d 0.05   e 0.35   f 0.8

g 0.32   h 0.525   i 0.71

5 Put these lists of numbers in order, starting with the smallest.

a 5.21   2.56   5.02   5.19   2.5

b 0.36   0.632   0.236   0.365   0.635

c 3.54   4.26   3.504   3.453   3.624

d 7.32   7.032   7.317   7.3   7.02

1 Round these numbers to the given degree of accuracy.

  a 46   (nearest 10)

  b 67.945   (2 decimal places)

  c 356   (nearest 100)

  d 13.46   (1 decimal place)

  e 34.56   (nearest 10)

  f 13.598   (2 decimal places)

2 Estimate the answer to these calculations:

  a $\dfrac{2.75 \times 6.3}{8.8}$

  b $\dfrac{11.7 - 7.6}{1.7 + 6.25}$

3 Use a written method to work out these.

  a 34.6 + 54.3   **b** 12.45 + 3.32   **c** 17.34 + 3.25

  d 15.7 − 3.6   **e** 3.75 − 2.24   **f** 8.78 − 3.46

  g 123.4 + 4.8   **h** 34.68 + 54.9   **i** 35.67 − 18.7

4 Use a mental or written method to solve each of these problems.

  a A bucket full of sand weighs 17.25 kg. The bucket weighs 1.7 kg. How much does the sand weigh?

  b James is saving his money in his money box. In July he has £34.56. In August he has £47.95. How much money did James save in August?

  c Paula is training to run a marathon. During 4 days of training she runs the following distances: 16.5 miles, 18.95 miles, 8.3 miles and 24.47 miles. Work out the total distance she runs during the 4 days.

1 Use a written method to work out these.

   a 14×23 **b** 19×34 **c** 24×156

   d $164 \div 4$ **e** $196 \div 7$ **f** $345 \div 15$

   g 3.4×5 **h** 14×4.5 **i** $85.61 \div 7$

2 Use a mental or written method to solve each of these problems.

   a 1 litre of petrol costs 90.4p.
      How much does it cost for 12 litres of petrol?

   b Sonia buys 32 greeting cards that cost £12.72.
      How much does each card cost?

   c Lawson drinks 1.55 litres of milk every day.
      How much milk does he drink in a week?

3 Put these numbers in order of size.
  Start with the smallest.

   a $\frac{1}{3}, \frac{3}{5}, 35\%, 0.3$ **b** $\frac{2}{3}, 60\%, 0.62, \frac{3}{7}$

   c $0.12, 24\%, \frac{1}{24}, \frac{1}{12}$ **d** $0.9, \frac{19}{20}, 0.89, \frac{11}{12}$

4 Simplify these fractions.

   a $\frac{4}{10}$ **b** $\frac{25}{45}$ **c** $\frac{16}{18}$

   d $\frac{9}{54}$ **e** $\frac{28}{32}$ **f** $\frac{56}{64}$

5 **a** Sam sends 137 text messages at 9p each. Work out an approximate cost for the text messages.

   b 6 shirts cost £159. Work out an approximate cost for each shirt.

   c The exchange rate for pounds to euros is £1 = €1.4. Work out approximately how many euros you would get for £300.

1 a Write these decimals as fractions in their simplest form.

    i 0.1   **ii** 0.08   **iii** 0.375

b Change these fractions to decimals without using a calculator.

    i $\dfrac{8}{10}$ **ii** $\dfrac{3}{4}$ **iii** $\dfrac{66}{300}$

c Change these fractions to decimals using a calculator. Give your answers to a suitable degree of accuracy.

    i $\dfrac{1}{13}$ **ii** $\dfrac{5}{9}$ **iii** $\dfrac{11}{17}$

2 Work out the value of each expression.

    a $3d + 2f$ when $d = 4$ and $f = 3$

    b $2p - 4q$ when $p = 15$ and $q = 4$

    c $5e - d + 4f$ when $e = 4$, $d = 2$ and $f = 6$

    d $5t - 6r - s$ when $t = 6$, $r = 1$ and $s = -3$

3 Write down a suitable estimate for each calculation.

    a $19.5 + 23.5$ **b** $478 - 212$

    c 1.98×4.05 **d** $18.45 \div 2.76$

    e $\dfrac{3.5 \times 9.7}{3.87}$ **f** $\dfrac{19.7 \times 21.3}{39.6}$

    g $\dfrac{19.45 \times 3.14}{2.1}$ **h** $\dfrac{199.3 \times 1.98}{9.67}$

4 39 548 people watch a football match.

    a Write 39 548 to the nearest thousand.

    b Write the value of the 3 in the number 39 548.

A2 HW2 Coordinates and function machine

1 Copy the coordinate grid onto square grid paper. Write the coordinates of the points shown.

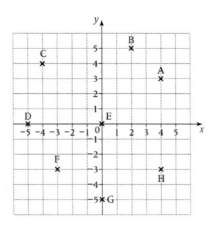

2 ABCD is a parallelogram.
A is the point (0, 3).
C is the point (0, −3).
Write the coordinates of vertices B and D.

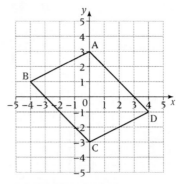

3 Copy and complete the input/output table for this function machine.

→ ×4 → +2 →

Input	Output
0	
1	
2	
3	
4	
	42

4 Draw a function machine like that in question **3** for each equation. Use your machines to calculate the outputs.

   a $y = 4x + 5$ **b** $y = x + 10$ **c** $y = \frac{x}{2} + 6$

   d $y = x - 5$ **e** $y = \frac{x}{4} + 5$ **f** $y = -2x - 6$

1 Complete the table of values for each equation.

a $y = 6x$

x	0	2	3	5	7
y					

b $y = 2x + 5$

x	2	3	5	7	10
y					

2 Match an equation in set A to its table of values in set B.

A

a $x + 2y = 10$ **i**

b $y = 2x - 4$ **ii**

c $y - x = 3$ **iii**

B

x	-1	2	3	5	7
y	-6	0	2	6	10

x	-2	0	2	4	6
y	6	5	4	3	2

x	-4	-2	1	2	4
y	-1	1	4	5	7

3 a Copy and complete this table of values for the equation $y = 2x - 2$.

x	-4	-2	0	2	6
y	-10				

b Write a list of coordinate pairs from your table.

c Copy the grid and plot the points from part **b**.
Join them with a straight line.

d From your graph, find the value of y when $x = 4$.

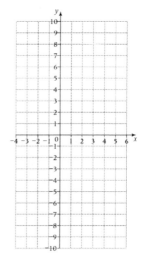

4 These coordinates lie on a straight line
(-2, 1) (-1, 2) (0, 3) (1, 4) (2, 5) (3, 6)

Copy and complete these statements.
The y-value is always equal to ___ + the x-value.
The equation of the line is $y =$ ___.

1 Choose an equation for each straight line graph.

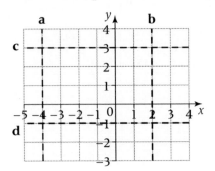

i $y = 3$
ii $x = 3$
iii $y = -4$
iv $x = 2$
v $y = 2$
vi $y = -1$
vii $x = -1$
viii $x = -4$

2 Copy the axes from question **1** onto squared paper.

  a Draw four lines to enclose a rectangle.

  b Write the equations of your four lines.

  c Write the coordinates of the vertices of your rectangle.

3 Here is a conversion graph for pounds to euros.
Use the graph to convert

  a €10 to pounds

  b €60 to pounds

  c £50 to euros

  d £75 to euros

  e €45 to pounds

  f €25 to pounds

  g £54 to euros

  h £9 to euros.

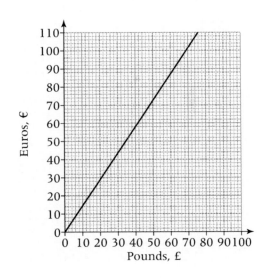

1 By using a conversion graph between °C and °F state which temperature is colder.

  a 10 °C or 10 °F

  b 25 °C or 50 °F

  c 0 °C or 30 °F

  d −25 °C or 25 °F

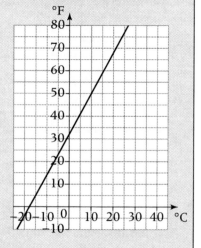

2 Complete the table of values for $y = 3x - 1$.

x	−1	0	1	2	3
y			2		

  a On a copy of the grid, draw the graph of $y = 3x - 1$.

  b Use your graph to find

    i the value of y when $x = -1.5$

    ii the value of x when $y = 2.3$

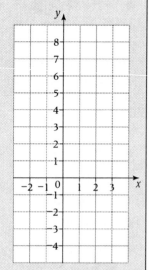

3 Draw the graph of $y = 4x - 8$.

1 Use an appropriate method of calculation to work out each of these.

    a $128 \div 8$ **b** $126 \div 9$ **c** $208 \div 13$

    d 12.5×8 **e** 2.15×12 **f** 15.4×15

    g $74 \div 5$ **h** $91.2 \div 6$ **i** $211.5 \div 9$

2 Complete the table of equivalent fractions, decimals and percentages, writing fractions in their simplest form.

Fraction	Decimal	Percentage
	0.5	
$\frac{1}{4}$		
		75%
$\frac{1}{10}$		
	0.125	

3 Write down the largest fraction in each pair.

    a $\frac{1}{2}$ $\frac{2}{5}$ **b** $\frac{7}{10}$ $\frac{9}{20}$

    c $\frac{1}{3}$ $\frac{4}{9}$ **d** $\frac{3}{4}$ $\frac{2}{7}$

    e $\frac{5}{12}$ $\frac{6}{11}$ **f** $\frac{13}{100}$ $\frac{1}{10}$

4 Simplify these expressions.

    a $m + m + m - m$ **b** $3k + 5k - k$

    c $4e - 4e$ **d** $17x + 12x + x$

1 Use your calculator, where necessary, to work out these.

a 4^2

b 12^2

c 4^3

d 2^3

e 17^2

f $(-2)^2$

g 7^3

h 15^2

i 12^3

j $12^2 + 4^3$

k $13^2 - 4^3$

l $5^3 - 9^2$

m $13^2 + 14^2 + 15^2$

n $3^3 + 4^3 + 5^3$

o $13^3 - 12^2 - 11$

2 Some numbers can be represented as the sum of two squares. For example:
$$4^2 + 5^2 = 16 + 25 = 41$$

Find all the numbers between 50 and 100 that can be represented as the sum of two square numbers.

3 Use your calculator, where necessary, to work out these. Give your answer to a suitable degree of accuracy.

a $\sqrt{81}$

b $\sqrt{119}$

c $\sqrt[3]{8}$

d $\sqrt[3]{1728}$

e $\sqrt{784}$

f $\sqrt[3]{729}$

g $\sqrt{50}$

h $\sqrt[3]{120}$

i $\sqrt{12}$

j $\sqrt{250}$

k $\sqrt[3]{10\,000}$

l $\sqrt{101}$

4 **a** A cube has a volume of 512 cm^3. What are its dimensions?

b A square has an area of 150 cm^2. What is the length of the square to 1 decimal place?

c Between which two whole numbers does $\sqrt{55}$ lie? Explain your answer.

d Between which two whole numbers does $\sqrt[3]{300}$ lie? Explain your answer.

N5 HW3 Powers of 10, factors and multiples

1 Use your calculator to work out these questions. In each case copy the question and fill in the missing numbers.

a $2^? = 8$ b $3^? = 27$ c $6^? = 36$
d $5^? = 125$ e $4^? = 64$ f $10^? = 1000$
g $2^? = 32$ h $3^? = 729$ i $2^? = 1$

2 Calculate each of these.

a 1.4×10 b $4.5 \times 10\,000$ c $766 \div 10$
d 4.5×10^3 e $54 \div 100$ f $23.9 \div 10^2$
g 2.05×10^2 h $35.6 \div 10^3$ i 0.345×10^4

3 **a** From the numbers in the oval, write the common factors of 36 and 48.

b From the numbers in the oval, write the common multiples of 6 and 9.

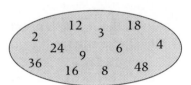

4 Here is a list of numbers.

15, 16, 17, 18, 19, 20, 21, 22, 23, 24, 25

From the list, write

a the two numbers that are multiples of 6
b the two numbers that are factors of 72.

5 **a** List all the factors of 36.
b List all the factors of 48.
c Which numbers are factors of both 36 and 48?

1 Find the highest common factor of

   a 6 and 8   **b** 12 and 36   **c** 15 and 18
   d 72 and 48   **e** 81 and 27   **f** 28 and 24

2 Find the least common multiple of

   a 3 and 4   **b** 7 and 6   **c** 9 and 5
   d 12 and 4   **e** 16 and 12   **f** 18 and 22

3 Using only numbers in the cloud, write

   a all the multiples of 4
   b all the square numbers
   c all the factors of 12
   d all the cube numbers
   e all the prime numbers.

> 64   8   16   1
> 12   48   20   2
> 28   100   216   3

4 Using only numbers in the circle write

   a numbers divisible by 3
   b common factors of 24 and 36
   c multiples of 9 and 6
   d multiples of 11
   e numbers that are cube and square numbers
   f prime numbers.
   g Three of the numbers in the circle are the square root of three other numbers in the circle. Write all 6 numbers.

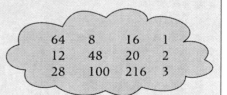

> 3
> 6
> 8   9   18
> 12
> 64   121
> 90
> 198   48
> 11
> 16   72

5 Work out

   a 6^2 **b** 10^4 **c** $(-1)^5$
   d $\sqrt{121}$ **e** $\sqrt[3]{1000}$ **f** $\sqrt[3]{27}$

1 a Copy and complete the table of values for $y = 2x + 3$.

x	0	1	2	3	4	5
y						

b On a copy of this grid, draw the graph of $y = 2x + 3$.

c Draw the lines $y = 4$ and $x = 3$ on your grid.

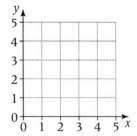

2 Simplify these expressions.

    a $2x + 4y + x$ **b** $3x - y + 4y$

    c $x - y + 3x$ **d** $4x - 4y + z - 3x - x + y$

    e $5x + 3y - 5x - 3y + x$ **f** $8x - 5y + 3x - 4y$

    g $x + y - x - y$ **h** $x + 3y - z - y + 2z$

3 From the numbers inside the oval, write the numbers that are

    a prime numbers

    b square numbers

    c cube numbers

    d factors of 24

    e prime factors of 12

    f highest common factor of 24 and 32

    g least common multiple of 8 and 6.

Numbers inside the oval: 2, 24, 144, 3, 24, 1, 49, 6, 36, 16, 8, 4, 216, 100, 9, 5, 1000

1 Write the next three terms in each sequence.
Give the rule that you used.

   a 4, 7, 10, 13, ...
   b 12, 15, 18, 21, 24, ...
   c 1, 1.5, 2, 2.5, 3, ...
   d 20, 17, 14, 11, ...
   e 0.1, 0.2, 0.3, 0.4, 0.5, ...
   f −7, −3, 1, ...

2 The start of a sequence is 1, 2, 4, ▉▉▉▉▉▉
The rest of the sequence is hidden!

   a Write what the next three terms could be and the rule you
have used.
   b Extend the sequence using a different rule.

3 Find the missing term in these sequences and write the
term-to-term rule.

   a 12, 15, 18, 21, __, __, __
   b 5, 9, __, 17, __, __
   c 10, 8, __, 4, 2, __, __
   d 7, __, 19, 25, __, __, __
   e __, __, 5, 11, __, 23, __
   f 35, 26, __, 8, __, __
   g 1, 1.5, __, __, 3, __, __

4 Generate the first five terms of a sequence with nth term

a $n + 4$	**b** $2n$	**c** n^2
d $2n + 1$	**e** $10 - n$	**f** $3n + 1$
g $4n - 2$	**h** $20 - 2n$	**i** $-3n$

1 Describe these sequences by giving the start number and the term-to-term rule.

   a 20 000, 2000, 200, 20, 2, ...

   b 5, 25, 125, 625, 3125, ...

   c 120, 60, 30, 15, 7.5, ...

   d 4, 8, 16, 32, 64, ...

   e 1, 0.1, 0.01, 0.001, 0.0001, ...

   f 10, −20, 40, −80, 160, ...

2 Describe these sequences in words by comparing them to the multiples of 3.

   a 4, 7, 10, 13, 16, ...

   b −3, −6, −9, −12, −15 ...

   c 1, 4, 7, 10, 13, 16, ...

   d 1.5, 3, 4.5, 6, 7.5, ...

3 Here are some patterns made from toothpicks.

Pattern 1   Pattern 2   Pattern 3

   a Draw pattern number 4.

   b How many toothpicks are used in pattern 10?

   c Write down a formula for the rule to work out the number of toothpicks needed for the nth pattern.

1 Here is a sequence.

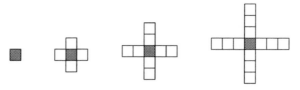

    a Write down the number of black squares and white squares for the next two patterns.

    b Complete the table for the number of white squares.

Pattern number	1	2	3	4	5
Number of white square			8		

    c Copy and complete this rule for the white squares.
       number of white squares = ___ × pattern number

    d Use your answers to parts **a** and **c** to write a rule for the total number of squares.

    e Use your rule from part **d** to work out the total number of squares in pattern number 10.

2 Repeat question 1 using this sequence of dots instead of squares.

3 Repeat question 1 using this sequence of triangles instead of squares.
How many triangles will there be in the 8th pattern?

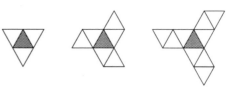

4 Repeat question 1 using your own pattern. Draw the first 3 patterns.

1 Here are the first five terms of a number sequence.

6   11   16   21   26   31

a Write down the next two terms of the sequence.
b Explain how you found your answer.
c Explain why 217 is not a term of the sequence.

2 Write the first five terms of the sequence with nth term

a $3n + 5$ **b** $8n - 4$ **c** $20 - 2n$
d $n^2 + 3$ **e** $12n - 5$ **f** $-2n + 5$
g $n^2 - 10$ **h** $30 - n^2$

3 Here is a pattern of hexagons.

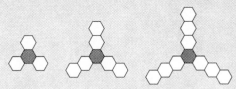

a Copy and complete the table.

Pattern number	1	2	3	4	5	n
Term						

b Work out how many hexagons there will be in the 10th pattern.

4 Here is a sequence made from dots.

The general term for the sequence is $D = 3n + 1$ where D is the number of dots and n is the pattern number.

a Work out the number of dots in pattern number 8.
b Which pattern has 31 dots?
c Is there a pattern with 67 dots? Explain your answer.

1 Write the equation of each line.

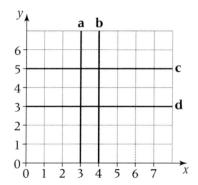

2 Here is a list of numbers.

3, 4, 6, 8, 12, 18, 24, 30, 36, 48, 72

Write the numbers which are

a factors of 12
b multiples of 6
c factors of both 12 and 8
d multiples of both 6 and 8
e factors of 72
f highest factor of 72.

3 Calculate these percentages, giving your answer to two decimal places where appropriate.

a 35% of £315 **b** 62% of 256 litres
c 18.3% of 878 miles **d** 6.8% of 552 m

4 Write the next 3 terms and the rule for each sequence.

a 2, 5, 8, 11, 14, ...

b 10, 8, 6, 4, 2, ...

c $\frac{1}{2}, \frac{3}{4}, 1, 1\frac{1}{4}, ...$

d 0.75, 0.8, 0.85, 0.9, ...

e 0.12, 0.11, 0.1, 0.09, ...

1 Sonia worked out the cost of petrol used in her car.

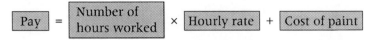

    a What is the cost of travelling one mile?
    b Sonia needs to travel 150 miles. Work out the cost of the petrol used.

2 Paul works as a painter. His hourly rate of pay is £20.
He calculates his pay using the formula:

$$\boxed{\text{Pay}} = \boxed{\text{Number of hours worked}} \times \boxed{\text{Hourly rate}} + \boxed{\text{Cost of paint}}$$

One week Paul works 36 hours and uses £126 worth of paint. Work out his pay for the week.

3 To exchange pounds for euros at the travel agent you have to subtract a £10 charge then multiply what is left by 1.4 Use this information to write a formula connecting number of euros, e, to number of pounds, p.

4 The cost of hiring a cement mixer is £29 plus £5 per hour.

    a Write a formula for how much it costs, C, to hire the cement mixer for n hours.
    b Use your formula to work out how much it costs to hire a cement mixer for 8 hours.

5 The rule for working out the perimeter of a square is to multiply the length of the square by 4.

    a Write a formula for the perimeter of the square.
    b Use the formula to work out the perimeter of a square with length

       i 6 cm   **ii** 8 cm   **iii** 12 cm.

1 For each formula, work out the value of y when $a = 3$ and $b = -2$.

a $y = 5a - b$	**b** $y = 2a - 3b$	**c** $y = ab$
d $y = a^2 + b$	**e** $y = 2ab^2$	**f** $y = b^3$
g $y = 2a^2 + b$	**h** $y = 3a - b$	**i** $y = b - a$

2 For each formula, work out the value of m, when $n = 3$, $p = -2$, $r = 5$

a $m = 2n + p$ **b** $m = r^2 + 4n$

c $m = \dfrac{np}{2}$ **d** $m = \dfrac{2rn}{6}$

e $m = \dfrac{4pr}{8}$ **f** $m = 2rn^2 + p$

g $m = 3p^2 + n$ **h** $m = \dfrac{2rn^2}{nr}$

3 Draw a conversion graph to convert these measurements.

a 4.5 feet to inches
b 28 inches to feet
c 13 feet to inches
d 78 inches to feet

> Hint:
> 12 inches = 1 foot

4 a Draw a conversion graph to convert pints to litres using
1 litre = $1\frac{3}{4}$ pints, 5 pints = $8\frac{3}{4}$ litres and 10 pints = $17\frac{1}{2}$ litres.

b Use the conversion graph to convert

i 8 pints to litres **ii** 12 litres to pints
iii 18 pints to litres **iv** 18 litres to pints.

1 The graph shows Mick's bike ride to a local park.

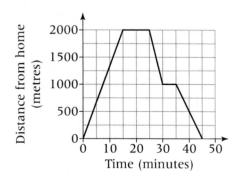

a How long did it take him to get there?

b How long did he spend at the park?

c What happened on the way home?

d Which part of the journey was he travelling the slowest? How can you tell?

2 The distance–time graph illustrates Derek's journey from his home to Mansfield and back.

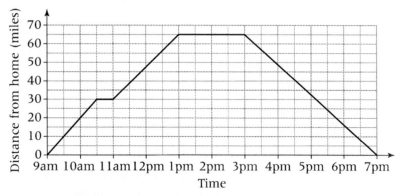

a What happened at 10.30 am?

b How long did he stop when he was 65 miles from home?

c How long did it take to travel home?

d Using the conversion of:

80 miles = 128 kilometres

convert the distance in miles to km.

e Using the formula speed = $\dfrac{\text{distance}}{\text{time}}$

work out Derek's speed in miles per hour between 11 am and 1 pm.

1 The graph shows Paula's journey to the shops.

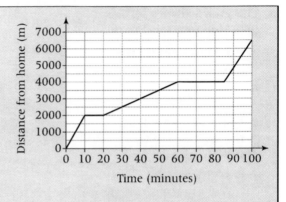

Distance from home (m) vs Time (minutes)

a How long did it take to get to the shops?

b How many times did she stop on the way?

c Using the formula $\text{speed} = \dfrac{\text{distance}}{\text{time}}$ work out the speed in metres per minute between

  i 0 and 10 minutes

  ii 20 and 60 minutes

  iii 85 and 100 minutes.

2 Here is a distance–time graph of David's journey from his house to the shops and back.

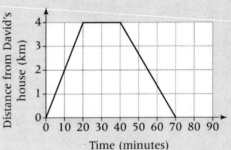

Distance from David's house (km) vs Time (minutes)

a How long did David stay at the shops?

b How long did it take David to walk home from the shops?

c Using the formula in question **1c**, find out how fast David walked home from the shops.

3 Teresa's sweets cost 6p each. A packet of crisps costs 25p.

a Write a formula for total cost of n sweets and c packets of crisps.

b If $n = 20$ and $c = 3$, use your formula to work out the total cost of sweets and crisps.

1 **a** Copy this coordinate grid.
Plot these coordinates
(4, 2), (−2, 3), (−4, 2)

b Find the fourth coordinate to
make a kite.

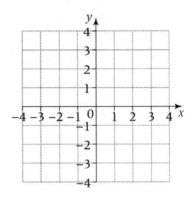

2 **a** Here is a sequence of
numbers.

14   19   24   29   34

i Write the next two terms of this sequence.

ii Is 96 a term of this sequence?
Explain how you know.

iii What is the 10th term of the sequence?

b Here is a sequence of numbers.

26   30   34   38   42

i Write the next two terms of this sequence.

ii Is 125 a term in this sequence?
Explain how you know.

iii Write down the first term in the sequence greater
than 100.

3 Use your calculator to work out each of these

a 18^3 **b** 9^5 **c** 3^9 **d** 11^4 **e** 5^5

1 David works as a carpenter. He is paid by the hour. Copy and complete the table to help him work out his pay.

Hours worked	Pay in £
8 hours	£92
1 hour	
12.5 hours	
35 hours	
42 hours	

2 **a** A recipe for fish pie used 500 g of fish for 4 people. How much fish is needed for a fish pie for

   i 1 person **ii** 6 people **iii** 10 people **iv** 14 people?

   b A building contractor pays nine workers £325.80 for a day's work. How much would he pay

   i 1 worker **ii** 4 workers **iii** 11 workers **iv** 28 workers?

   c There are 30.8 lb in 14 kg. How many lb are these in

   i 1 kg **ii** 1000 kg **iii** $\frac{1}{4}$ kg **iv** 1.5 kg?

3 Write these ratios in the form $1 : n$.

   a 3 : 9 **b** 4 : 16 **c** 5 : 10 **d** 7 : 21
   e 10 : 25 **f** 15 : 60 **g** 4 : 28 **h** 6 : 54
   i 22 : 33 **j** 18 : 24 **k** 27 : 45 **l** 36 : 48

4 In each of these questions work out the rate of petrol consumption for each person's car.

	Distance travelled (miles)	Petrol used (litres)	Rate of petrol consumption (miles per litre)
Mr Milligan	400	40	
Mrs Tomes	360	30	
Mr Worrall	950	80	
Miss Turpin	725	50	
Mr Edwards	103.5	11.5	

1 The exchange rate for pounds to euros is approximately
£1 = €1.40.

  a Calculate how many euros you get for

    i £10
    ii £15
    iii £120
    iv £500

  b Calculate how many pounds you get for

    i €5.60
    ii €154
    iii €168
    iv €224

2 **a** To make mortar you mix sand and cement in the ratio
5 : 2. How much cement is needed for 20 kg of sand?

  b In a school the ratio of staff to pupils is 2 : 45. If there are
450 pupils in the school, how many staff are there?

  c A scale on a map is 1 : 250. If the distance on the map is
5 cm what is the distance in real life?

  d A model car has a scale of 1 : 15. What is the length of the
model car if the real life car measures 4.5 metres?

Example

Divide £28 in the ratio 3 : 4.

$3 + 4 = 7$
$28 \div 7 = 4$
$4 \times 3 = 12, 4 \times 4 = 16$
Answer: £12 and £16

3 Solve each of these problems.

  a Divide £20 in the ratio 1 : 4.
  b Divide £90 in the ratio 2 : 7.
  d Divide 72 kg in the ratio 3 : 5.
  e Divide 125p in the ratio 13 : 12.
  f Divide 84p in the ratio 5 : 2.

1 Write these ratios in the form $1:n$.

    a $3:12$ **b** $2:18$ **c** $9:45$ **d** $14:28$

    e $5:35$ **f** $15:75$ **g** $18:24$ **h** $36:18$

2 Solve each of these problems.

    a Divide £25 in the ratio $3:2$

    b Divide £120 in the ratio $1:5$

    c Divide 144 kg in the ratio $5:7$

    d Divide 240p in the ratio $7:3$

    e Divide 132p in the ratio $5:6$

3 Omar buys petrol from his local garage.

On Sunday he fills up his tank.

On Monday, his tank was $\frac{3}{5}$ full.

    a What fraction of the full tank of petrol has been used?

    b Write $\frac{3}{5}$ as a decimal.

    c Write $\frac{3}{5}$ as a percentage.

The garage gives the approximate conversion
from gallons to litres

    1 gallon = 4.5 litres

    d Use the conversion to convert

      i 3 gallons to litres

      ii 4.5 gallons to litres.

4 This is a list of ingredients for making apple and blackberry crumble for 4 people.
Work out the amount of each ingredient needed to make apple and blackberry crumble for 10 people.

Ingredients for 4 people:

  80 g plain flour
  90 g soft brown sugar
  60 g butter
 100 g blackberries
    4 cooking apples

1 Write the next three terms in each of these sequences.

   a 3, 5, 7, 9, ...
   b 35, 32, 29, 26, ...
   c 7, 13, 19, 25, 31, ...
   d 10, 6, 2, −2, −6, ...
   e −14, −11, −8, −5, ...

2 Describe in words the rule for each of the sequences in question **1**.

3 Work out:

a $3 - 7$	**b** $-4 + 2$	**c** $-6 - 3$
d $8 - -2$	**e** $-5 - 2$	**f** $-6 + 4$
g $-7 - -1$	**h** $-4 - -7$	**i** $-4 + 9$

4 Work out:

a 6×7	**b** 4×3	**c** 8×3
d 5×9	**e** 6×8	**f** 4×8
g 9×2	**h** 5×6	**i** 3×5

5 Work out the missing values:

   a $4 + \square = 11$ **b** $9 + \square = 18$
   c $10 - \square = 7$ **d** $4 - \square = 1$
   e $6 \times \square = 24$ **f** $\square \times 3 = 21$

1 Work out the outputs for these function machines.

a 6 → ×8 →

b 8 → ×6 →

c 11 → +7 →

d 7 → −11 →

e 4 → ×6 →

f 75 → ÷5 →

g 17 → −9 →

h 4 → +26 →

2 Draw the inverse machines for these function machines.

a 12 → −7 → 5

b 0.5 → ×4 → 2

c 28 → ÷2 → 14

d −4 → ×5 → −20

3 Use function machines to solve these 'think of a number problems'.

a I think of a number and times it by 2. The answer is 16. What number did I think of?

b I think of a number and subtract 7. The answer is 7. What number did I think of?

c I think of a number and add −2. The answer is 10. What number did I think of?

4 Use function machines to solve these equations.

a $a + 2 = 10$ **b** $x + 4 = 11$ **c** $n - 5 = 11$

d $s + 9 = 24$ **e** $r - 10 = 10$ **f** $9 + f = 10$

g $a - 4 = 17$ **h** $x - 4 = -1$ **i** $s - 5 = 9$

j $n + 5 = -9$ **k** $a + 1 = 1$ **l** $q + 7 = 13$

A5 HW3　Solving equations

1 Draw function machines for these equations.
Use inverse machines to solve them.

a $6x = 24$　　　　**b** $9y = 27$　　　　**c** $4z = 44$
d $7r = 21$　　　　**e** $8s = 48$　　　　**f** $4t = 16$
g $\frac{x}{5} = 4$　　　　**h** $\frac{y}{3} = 8$　　　　**i** $\frac{v}{5} = 32$
j $\frac{w}{4} = 10$　　　**k** $\frac{t}{10} = 10$　　**l** $\frac{s}{6} = 12$

2 a Match each function machine in set A with an equation in Set B.

Set A

a $x \rightarrow \boxed{+6} \rightarrow 9$　　**b** $x \rightarrow \boxed{\div 5} \rightarrow 10$　　**c** $x \rightarrow \boxed{+9} \rightarrow 2$

d $x \rightarrow \boxed{\times 4} \rightarrow 16$　　**e** $x \rightarrow \boxed{-10} \rightarrow -4$　　**f** $x \rightarrow \boxed{\div 7} \rightarrow 8$

g $x \rightarrow \boxed{-4} \rightarrow 10$　　**h** $x \rightarrow \boxed{\times 9} \rightarrow 18$

Set B

i $x - 4 = 10$　　**ii** $\frac{x}{5} = 10$　　**iii** $9x = 18$　　**iv** $x - 10 = -4$
v $x + 6 = 9$　　**vi** $4x = 16$　　**vii** $\frac{x}{7} = 8$　　**viii** $x + 9 = 2$

b Solve each of the equations.

3 Solve these equations.

a $x + 4 = 8$　　　**b** $x - 6 = -1$　　　**c** $x + 7 = 11$
d $x - 7 = -2$　　　**e** $10 + x = 12$　　**f** $8 - x = 5$
g $14 - x = -1$　　**h** $17 + x = 15$

4 a Write an equation for each of these.

　i a number added to 7 equals 19
　ii a number subtracted from 10 equals 3
　iii when subtracting 4 from a number you get 12
　iv a number added to 8 equals 2
　v a number added to −2 makes 1

b Solve each of your equations.

1 Solve these equations.

    a $3x = 12$ **b** $4x = 20$ **c** $\frac{x}{3} = 5$

    d $20 = 10x$ **e** $\frac{x}{6} = 4$ **f** $\frac{h}{6} = 4$

    g $5x = -10$ **h** $\frac{t}{2} = 15$ **i** $2x = -6$

2 If $x = 8$, pair the cards together to make equations that are balanced.

    a $\boxed{x + 2}$ **b** $\boxed{\frac{x}{4}}$ **i** $\boxed{10}$ **ii** $\boxed{1}$

    c $\boxed{9 - x}$ **d** $\boxed{3x}$ **iii** $\boxed{6}$ **iv** $\boxed{24}$

    e $\boxed{x - 2}$ **f** $\boxed{\frac{40}{x}}$ **v** $\boxed{2}$ **vi** $\boxed{5}$

3 Sharanjit has a bag with some sweets in it.
She eats 5 sweets and counts the number of
sweets she has left.
She has 9 sweets left.

    a Write an equation to show this information.
    b Solve the equation to work out how many sweets
       Sharanjit had at the beginning.

4 A builder charges £15 an hour.

    a Write a formula to work out how much he would
       charge, c, for working n hours.
    b If the builder works for 5 hours, use your formula to
       work out how much he will charge.
    c If the builder charged £90, use the inverse to work out
       how many hours he worked.

1 Copy the grid.

a Write the coordinates of point A.

b i On the grid, plot the point (−3, 1) and label it B.

ii On the grid, plot the point (−1, −4) and label it C.

c A, B and C are three corners of a square. Write the coordinates of D, the fourth corner of the square.

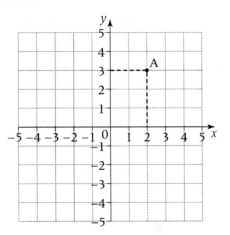

2 Estimate the size of each angle.

a   **b**   **c**   **d**

3 Measure these lines

i in millimetres

ii in centimetres

a _____

b _____

c _____

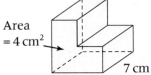

4 Using this formula work out the volume of these shapes.

Volume of prism	=	Area of cross-section	×	length

a

Area = 4 cm²

7 cm

b

6 cm

3 cm

5 cm

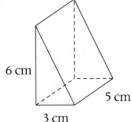

73

1 Convert these measurements to metres.

Hint: 10 mm = 1 cm, 100 cm = 1 m and 1000 m = 1 km

a 350 cm   **b** 15 km   **c** 2700 mm
d 55 cm   **e** 450 mm   **f** 3 km 200 cm

2 Convert these measurements to kilograms.

Hint: 1000 mg = 1 g, 1000 g = 1 kg and 1000 kg = 1 t

a 1 tonne   **b** 2500 g   **c** 2 500 000 mg
d 4.5 tonnes   **e** 14 kg 200 g **f** 500 g 200 mg

3 Copy and complete each **distance** conversion.

Hint: 5 miles ≈ 8 kilometres

a ___ miles = 8 kilometres   **b** 7.5 miles = ___ kilometres
c ___ miles = 24 kilometres   **d** ___ miles = 20 kilometres

4 Copy and complete each **mass** conversion.

Hint: 1 kg ≈ 2.2 lb and 30 g ≈ 1 oz

a 3 kg = ___ lb   **b** ___ kg = 8.8 lb
c 9 kg = ___ lb   **d** ___ kg = 1.1 lb
e 60 g = ___ oz   **f** ___ g = 7 oz
g 45 g = ___ oz   **h** ___ g = 4.5 oz

5 A farmer converts litres to gallons using this diagram.

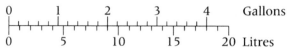

Use the diagram above and convert

a 10 litres to gallons   **b** 2 gallons to litres
c 3.5 gallons to litres   **d** 15 litres to gallons.

1 Find the perimeter of each shape.

a

5 cm

Equilateral triangle

b 4 cm

Square

c 3 cm

Regular hexagon

2 The perimeter of each rectangle is 24 cm.
Find the missing lengths.

a 8 cm

? cm

b 6 cm

? cm

c 9 cm

? cm

3 Work out the area of each rectangle in question **2**.
State the units of your answers.
Area = length × width

4 Find the area of each rectangle.

a 4 cm

2 cm

b 3 cm

3 cm

c 6 cm

2.5 cm

d 12 cm

9 cm

1 Find the missing lengths. Give the units of your answers.

a

3 cm | ? cm
| Area = 36 cm²

b

2.5 cm | ? cm
| Area = 7.5 mm²

2 The diagram shows a rectangular carpet.

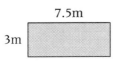

7.5m

3m

Work out the area of the carpet.

3 a Draw the net for this cuboid.
b Find the surface area of the cuboid.

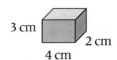

3 cm
2 cm
4 cm

4 Work out the surface area of these cuboids. State the units of your answers.

a
2 cm
2 cm
3 cm
3 cm

b
3 cm
2 cm
4 cm

c
15 mm
5 mm
10 mm

d
2.5 m
1 m
2 m

e
3 cm
2 cm
4.5 cm

f
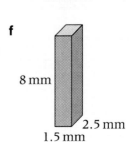
8 mm
2.5 mm
1.5 mm

1 Copy and complete each conversion.

Hint: Use the conversion hints on page 74 to help you.

a 45 cm = ___ mm   **b** 2300 cm = ___ m
c 3500 g = ___ kg   **d** 300 ml = ___ litres
e 4 tonnes = ___ kg   **f** 3.5 km = ___ m
g 12 000 cm = ___ m   **h** 0.6 litre = ___ ml

2 Put the measurements in order of size, starting with the smallest.

a 5 miles, 5 kilometres
b 45 mph, 80 km/h
c 11 inches, 31 centimetres, 1.5 feet
d 5 pints, 5 litres, 5 gallons
e 3 yards, 4 feet, 100 centimetres, 100 inches
f 15 kilograms, 20 lb, 10 oz

3 Work out the perimeter and area of each shape.
State the units of your answers.

a   5 cm   **b**   4 m   **c**

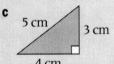

2.5 m

5 cm   3 cm

4 cm

4 Find the surface area of these cuboids.
State the units of your answers.

a   3 cm   2 cm   4 cm

b   4 cm   3 cm   3 cm

c   25 mm   5 mm   15 mm

1 Copy the grid.

a Write the coordinates of point A.

b i On the grid, plot the point (0, −2) and label it B.

ii On the grid, plot the point (1, 3) and label it C.

c A, B and C are three corners of a rectangle. Write the coordinates of D, the fourth corner of the rectangle.

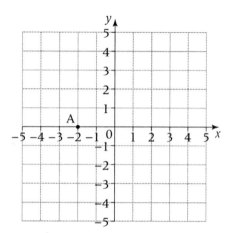

2 Copy and complete these conversions.

a 15 miles = ___ kilometres   **b** ___ miles = 36 kilometres
c 25 cm = ___ inches   **d** ___ kg = 6.6 lb
e 35 mph = ___ km/h   **f** ___ litres = 10 pints
g 30 cm = ___ feet   **h** 20 g = ___ oz
i 600 ml = ___ pints   **j** 5 kg = ___ lb

3 Calculate the area of each triangle. State the units of your answers.

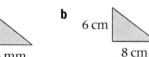

a 10 mm, 13 mm

b 6 cm, 8 cm

c 3 m

G2 HW2 Angles

1 **a** Draw a rectangle and label the perpendicular sides with ∟

b Draw a parallelogram and label the parallel sides with → and ↠

2 For each angle from **a** to **f** state

i your estimate of the angle in degrees

ii the measurement of the angle in degrees.

Set your answers out in a table like this one

Angle letter	Estimate of angle	Measurement of angle
a		

a

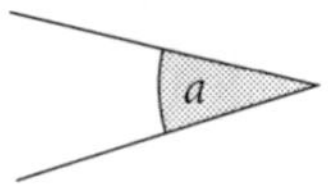

b

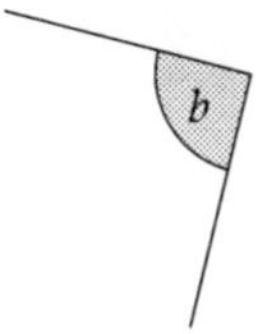

c

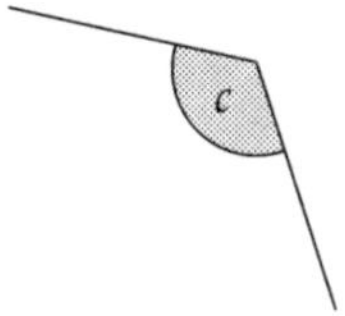

d

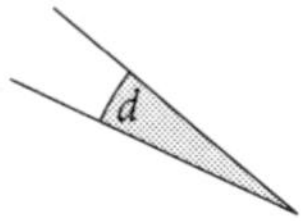

e

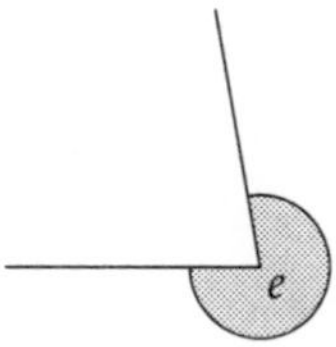

f

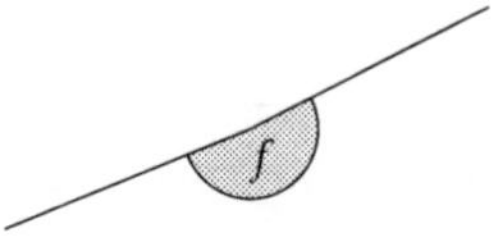

3 Calculate the size of the angles marked by letters in each diagram.

a

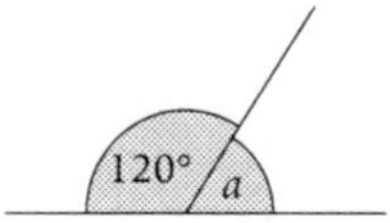

b

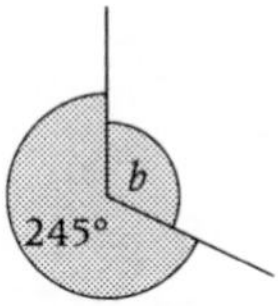

c

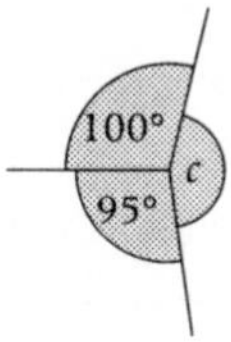

d

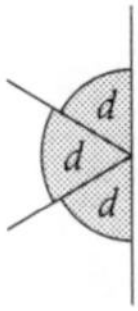

e

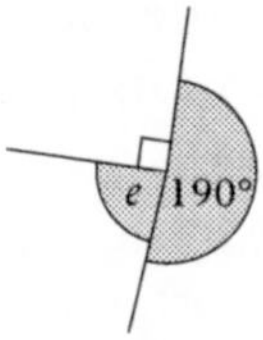

f

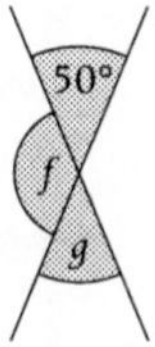

1 Calculate the size of the missing angles in each diagram.

a
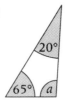

b

c
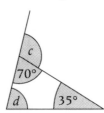

d
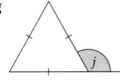

e

f
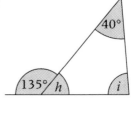

g

h

2 In the diagram QRS is a straight line.

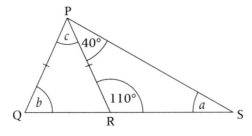

a i Work out the size of the angle marked *a*.
   ii Give reasons for your answer.

b i Work out the size of the angle marked *b*.
   ii Give reasons for your answer.

c i Work out the size of the angle marked *c*.
   ii Give reasons for your answer.

1. Measure and calculate the perimeter of these shapes in
   a cm
   b mm.

   i    ii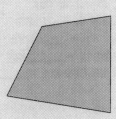

2. Measure each of these angles and state what type
   of angle it is.

   a   b

   c    d

3. Calculate the size of the missing angle in each diagram.

   a   b   c

   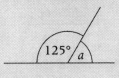   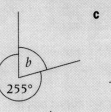

4. a i What is the size of
        angle ABC?
     ii Give a reason for your
        answer.
   b i What is the size of
        angle ADC?
     ii Give a reason for your
        answer.

   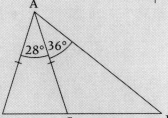

1 Copy and complete these conversions.

Hint: Use the conversion hints on page 74 to help you.

a ___ m = 340 cm   **b** 8 km = ___ m
c 1400 mm = ___ cm   **d** 2300 g = ___ kg
e 3400 ml = ___ litres   **f** 6.5 litres = ___ ml
g 5 tonnes = ___ kg   **h** 12 kg 300 g = ___ g
i 45 000 000 cm = ___ km   **j** 0.004 tonnes = ___ g

2 By using the conversion graph between stones and kilograms, state which weight is heavier.

a 4 stone or 4 kg
b 10 kg or 5 stone
c 10 stone or 70 kg
d 15 stone or 84 kg

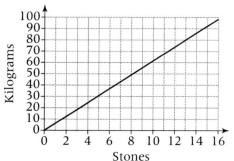

3 Find the missing area or length in each rectangle.

a

6 cm

2.5 cm   Area = ? cm²

b

8 cm

? cm   Area = 36 cm²

c A rectangle measuring 5.5 cm by 3 cm.
d A rectangle with area 32 cm² and one side 8 cm.

4 Draw and label these angles.

a 34°   **b** 125°   **c** 315°   **d** 95°

1 Copy these shapes and draw in all of the lines of symmetry.

a   b   c

d   e   g

2 Draw a shape that has

a exactly 1 line of symmetry
b exactly 2 lines of symmetry
c no lines of symmetry.

3 a Copy and complete the table.

Shape	Number of lines of symmetry	Order of rotational symmetry
Rectangle		
Regular octagon		
Parallelogram		
Regular pentagon		
Rhombus		

b Extend the table by adding in some other well known two-dimensional shapes.

1 Work out the size of the angles marked by letters.
Give a reason for each answer.

a

b
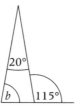

c
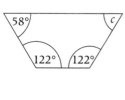

d
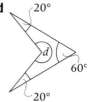

e
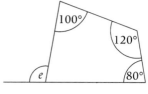

f
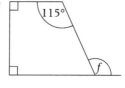

2 a Draw a circle with a radius of 3 cm.

  b Draw a chord of length 4 cm inside the circle.

  c Shade this segment of the circle.

3 Use a protractor and compasses to construct these sectors.

a
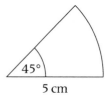

45°
5 cm

b

6 cm
125°

1 Write the mathematical names for these 2-D and 3-D shapes.

a 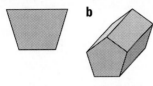 b c d e

f g h i j

2 Copy this net of a cube.
The net is folded to make a cube.
Two other vertices meet at A.

a Mark each of them with the letter A.

The length of each edge of the cube is 3 cm.

b Work out the volume of the cube.

A

3 Using this formula work out the volume of each shape.

Volume of prism = Area of cross-section × length

a

Area
= 5 cm²

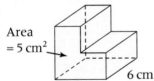

6 cm

b

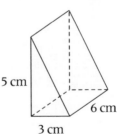

5 cm

6 cm

3 cm

1 Copy the diagrams and shade one square so that the shapes have exactly one line of symmetry.

a   **b**

2 Draw these solids on isometric paper after the shaded shape has been removed.

a 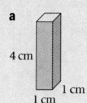 **b** 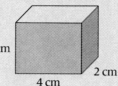 **c**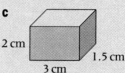

3 For each of these shapes, put a tick in the box if the statement is always true.

Statement	Square	Rhombus	Trapezium
Diagonals cross at right angles			
Opposite sides are parallel			
All angles are the same size			

4 a Work out the size of the missing angle, x.
b Give a reason for your answer.

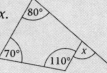

5 Work out the volume of each cuboid.

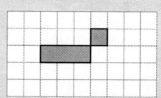

a 4 cm, 1 cm, 1 cm

b 3 cm, 4 cm, 2 cm

c 2 cm, 3 cm, 1.5 cm

1 Solve these equations.

   a $2x + 5 = 17$ **b** $3x - 6 = 3$ **c** $4x + 3 = 23$

   d $10 + 5x = 20$ **e** $7x - 2 = 19$ **f** $15x + 3 = 33$

2 Solve these equations.

   a $5x - 8 = 2$ **b** $\dfrac{3}{x} = 6$ **c** $20 - 3x = 14$

   d $3x = -12$ **e** $4 - x = 10$ **f** $4x + 3 = 13$

   g $\dfrac{x}{3} + 2 = 6$ **h** $10 - 3x = 1$ **i** $\dfrac{20}{x} = 2.5$

3 Complete the table of equivalent fractions, decimals and percentages, writing fractions in their simplest form.

Fraction	Decimal	Percentage
	0.2	
$\frac{3}{4}$		
		25%
$\frac{1}{100}$		
	0.375	

4 Calculate the size of the angles marked by letters in each diagram.

a

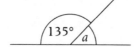

b

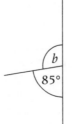

c

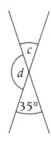

d

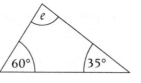

e

f

1 Round these numbers to the given degree of accuracy.

  a 45.9   (nearest 10)
  b 29 387   (nearest 100)
  c 493.4   (nearest whole number)
  d 34.57   (1 decimal place)
  e 12 098   (nearest 100)
  f 349.475 (2 decimal places)
  g 1256   (nearest 10)
  h 9.142   (nearest 10)
  i 189.96   (1 decimal place)
  j 12.9765 (2 decimal places)

2 Give a suitable estimate for each of these.
Show your workings clearly.

  a 6.95×3.21 **b** $19.7 \div 4.3$

  c $\dfrac{3.67 \times 5.24}{4.87}$ **d** $\dfrac{47.9 \times 32.1}{15.4}$

3 a Estimate the value of
$$\frac{19.6 \times 21.3}{(9.6)^2}$$

  b Use your calculator to work out the exact value giving
your answer to 2 decimal places.

4 a Janet buys 250 old postcards for 45p each. Work out the
cost of all of the postcards together.

  b Paul buys 14 tins of paint. The total cost of the tins of
paint is £103.60. Work out the cost of one tin of paint.

  c Teresa needs 215 tiles for a room.
Tiles are sold in boxes of 15.

    i What is the fewest number of boxes she needs to buy?
    ii Each box cost £7.80. What is the total cost of the boxes
of tiles Teresa needs to buy?

1 Work out each of these, changing your answer to its simplest form where necessary.

a $2 \times \dfrac{4}{7}$ **b** $\dfrac{4}{7} \times 3$ **c** $\dfrac{5}{9} \times 4$

d $\dfrac{5}{6} \times \dfrac{5}{8}$ **e** $\dfrac{3}{7} \times \dfrac{4}{9}$ **f** $\dfrac{2}{11} \times \dfrac{1}{4}$

g $\dfrac{1}{6} \times \dfrac{4}{9}$ **h** $\left(\dfrac{1}{6}\right)^3$ **i** $\left(\dfrac{3}{4}\right)^3$

2 Calculate these leaving your answers as fractions in their simplest form where necessary.

a What is the total weight of 4 crates each weighing $\dfrac{3}{4}$ kg?

b Adu takes $\dfrac{3}{5}$ of an hour to do each piece of homework. How long does it take him to do 6 pieces of homework?

c A rectangle is $\dfrac{3}{5}$ m long and $\dfrac{3}{7}$ m wide. What is the area of the rectangle?

3 a Work out

$$\dfrac{7}{8} \times \dfrac{5}{9}$$

b Estimate the value of

$$\dfrac{29 \times 602}{298}$$

4 In 2005, Lutterworth Hockey Club's total income from a lottery grant and members' fees was

Lottery grant   £350
Members' fees   60 at £45 each.

a Work out the total income of the club for 2005.

b Find the Lottery grant as a fraction of the club's total income for the year 2005. Give your answer in its simplest form.

5 Use your calculator to work out each of these. Write all the figures on your calculator display.

a $\dfrac{6.53^2 \times 2.19 + 7.34}{5.13 - 3.78}$

b $\dfrac{94.39 - (4.8 + 2.71)^2}{5.81^2 - 5.42}$

1 Stephen invests £50 in a bank account.
Simple interest of 4% is added at the end of each year.
Work out how much money Stephen has at the end of

　a the first year
　b three years.

2 a Estimate the value of
$$\frac{89.4 \times 34.5}{1.92 \times 30.4}$$

　b Use your calculator to work out the value of
$$\frac{12.35 \times (3.4 + 4.9)}{2.4^2 \times 1.3}$$

　Give your answer to 2 decimal places.

3 Calculate each of these, leaving your answer in its simplest form where necessary.

　a $2 \times \frac{2}{5}$　　　　**b** $\frac{3}{5} \times 4$　　　　**c** $\frac{4}{9} \times 3$

　d $\frac{2}{5} \times \frac{3}{10}$　　　**e** $\frac{4}{7} \times \frac{5}{6}$　　　**f** $\frac{4}{11} \times \frac{2}{3}$

　g $\frac{3}{5} \times \frac{7}{8}$　　　**h** $\left(\frac{2}{3}\right)^2$　　　**i** $\left(\frac{4}{5}\right)^2$

4 Calculate each of these amounts. Where appropriate round your answer to 2 decimal places.

　a $\frac{2}{15}$ of £630　　　**b** $\frac{2}{7}$ of 80 g

　c $\frac{7}{9}$ of 5470 m²　　**d** $\frac{8}{11}$ of 981 kg

1 Copy and complete each conversion.

   a 35 cm = ___ mm   **b** 3200 cm = ___ m

   c 4500 g = ___ kg   **d** 500 ml = ___ litres

   e 6 tonnes = ___ kg   **f** 4.5 km = ___ m

   g 18 000 cm = ___ m   **h** 0.7 litre = ___ ml

2 Name these shapes:

a   **b**   **c**

d    **e**   **f**

3 Write down the equations of these lines.

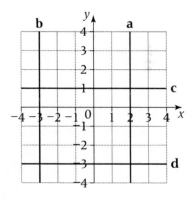

4 **a** Draw a circle with a radius of 6 cm.

   b Draw a chord of length 10 cm inside the circle.

   c Shade this segment of the circle.

1 Copy and complete the diagrams to show the reflections of the shapes in the mirror lines. The first one has been done for you as an example.

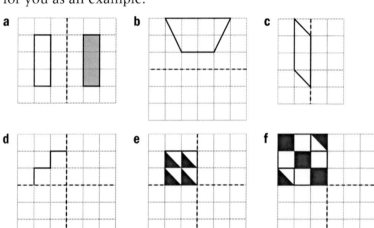

a   b   c

d   e   f

2 Copy the shapes and rotate them through the given angle and direction about the dot.

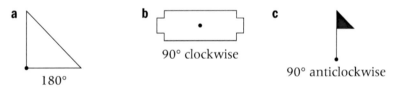

a

180°

b

90° clockwise

c

90° anticlockwise

3 State the angle and direction of turn for each of these rotations. The shaded shape is the start position.

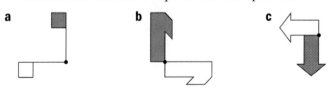

a   b   c

4 Give the angle and direction of a rotation that is identical to a rotation of

   a 90° anticlockwise   **b** 150° clockwise

   c 335° clockwise   **d** 45° anticlockwise.

1 Which shapes are translations of the shaded shape?

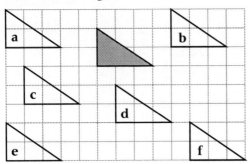

2 Describe the transformation of the shaded shape to each of the other triangles.

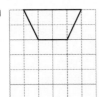
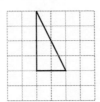

3 Copy these shapes onto square grid paper. Show how each shape tessellates, using rotations and translations.

a

b

c

4 Copy these shapes onto isometric paper. Show how each shape tessellates.

1 Copy and complete the diagrams to show the reflections of the shapes in the mirror lines.

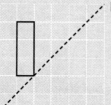 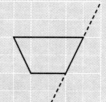 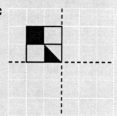

a   b   c

2 Which three of these regular polygons will tessellate?

Hexagon   Pentagon   Octagon   Square
Heptagon   Nonagon   Equilateral triangle

3 Copy these shapes and rotate them through the given angle and direction about the dot.

a   b   c

 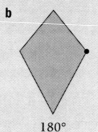 

90° anticlockwise

90° clockwise

180°

4 Describe these translations

  a A to B
  b A to C
  c A to D
  d B to A
  e C to D
  f D to A
  g D to B

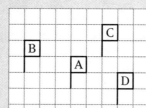

1 Solve these equations.

a $x + 8 = 20$　　**b** $15 - x = 5$

c $8x = 32$　　**d** $4x = -12$

e $14 + x = 13$　　**f** $x - 10 = 25$

g $\frac{x}{4} = 20$　　**h** $\frac{x}{2} = -10$

2 Use your calculator, where necessary, to work out these.

a 5^2　　**b** 11^2　　**c** 3^3

d 1^3　　**e** $(-3)^2$　　**f** 11^3

g $\sqrt{49}$　　**h** $\sqrt{169}$　　**i** $\sqrt[3]{27}$

j $\sqrt[3]{2197}$　　**k** $\sqrt{2500}$　　**l** $\sqrt[3]{1000}$

3 Copy the shapes and rotate them through the given angle and direction about the dot.

a

270° clockwise

b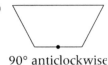

90° anticlockwise

c

180°

4 Reflect each shape in the mirror line.

a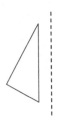

b

c

1 Copy and complete the diagrams to show the reflections of the shapes in the mirror line $y = 3$.

a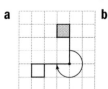

b

c

2 Give the equation of the mirror line for each reflection.

a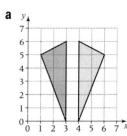

b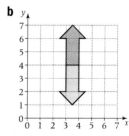

c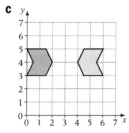

3 State the angle and direction for each of these rotations. The shaded shape is the starting position.

a 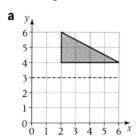   b 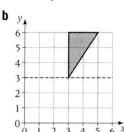   c 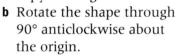   d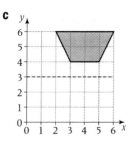

4 a Plot and join the points (0, 0) (2, 4) (−2, 4) on a copy of the grid.
  b Rotate the shape through 90° anticlockwise about the origin.
  c Give the coordinates of the rotated points.

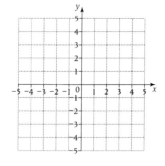

G5 HW3 Translations and enlargements

1 Describe these transformations.

    a A to B   **b** A to C
    c B to C   **d** B to D
    e C to D   **f** C to E
    g D to E   **h** D to A

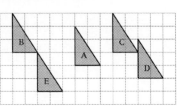

2 On square grid paper, draw the axes shown.

    a Plot and join the points (2, 1), (4, 1), (5, 4), (3, 4).
       Label it shape A. Name the shape.

    b Translate shape A by

       i $\begin{pmatrix} -5 \\ 1 \end{pmatrix}$ and label it B

       ii $\begin{pmatrix} -4 \\ -5 \end{pmatrix}$ and label it C

       iii $\begin{pmatrix} 1 \\ -4 \end{pmatrix}$ and label it D.

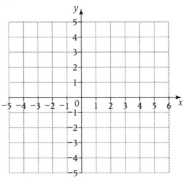

3 Calculate the scale factor of these enlargements.

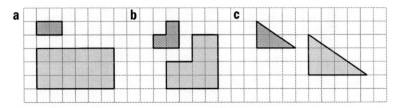

4 Decide if these rectangles are enlargements of the shaded rectangle. If so, calculate the scale factor.

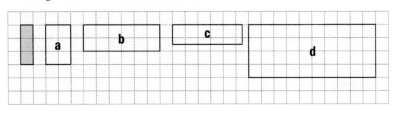

Use square grid paper for the enlargements on this page.

1 Enlarge each shape by the given scale factor.

a 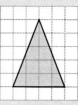   b    c   d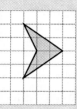

    scale factor 2   scale factor 3   scale factor 4   scale factor 2

2 Enlarge this shape by scale factor 2.

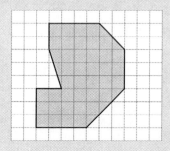

3 Copy the grid onto squared paper.

  a Reflect triangle A in the y-axis and label it B.

  b Translate triangle B by $\begin{pmatrix} 1 \\ 5 \end{pmatrix}$ and label it C.

  c Rotate triangle C through the origin (0, 0) by 90° anticlockwise and label it D.

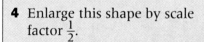

4 Enlarge this shape by scale factor $\frac{1}{2}$.

Hint: This means the shape will get smaller.

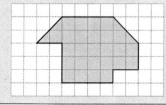

1 Solve these equations.

   a $x + 4 = 7$ **b** $10 - x = 4$

   c $3x = 12$ **d** $\frac{x}{2} = 10$

   e $6 + x = 12$ **f** $x - 6 = 15$

   g $4x = 16$ **h** $\frac{10}{x} = 5$

   i $x + 2 = -5$ **j** $4x = -20$

2 Solve these equations.

   a $x + 12 = 20$ **b** $10 - x = 7$

   c $6x = 24$ **d** $\frac{x}{3} = 27$

   e $13 + x = 22$ **f** $x - 7 = 15$

   g $3x = -9$ **h** $\frac{x}{2} = -6$

3 Copy the grid onto square grid paper.

   a Reflect triangle A in the
      x-axis.
      Label it B.

   b Translate triangle B by $\binom{6}{7}$.
      Label it C.

   c Rotate triangle C through
      the origin (0, 0) by 90°
      anticlockwise.
      Label it D.

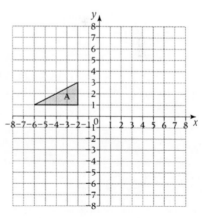

A6 HW2 Equations and function machines

1 Solve these equations using the balance method.

a $x + 5 = 13$ **b** $\frac{x}{4} = 12$ **c** $20 - x = 19$

d $3x = 15$ **e** $x + 5 = 4$ **f** $\frac{20}{x} = 2$

g $10 + x = 12$ **h** $15 - x = 16$

2 Write the different equations you can make using these numbers and symbols.
Solve each equation to find the value of x

a x 20   5   =   +   −
b x 12   4   =   ÷   ×

3 Draw function machines for these expressions.

a $2x - 3$ **b** $5n + 4$ **c** $2m - 4$
d $7p + 2$ **e** $5y + 1$ **f** $8z - 4$
g $3s - 11$ **h** $5t + 7$

4 Here are two function machines.

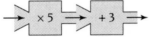 

Inputs and outputs for the two function machines have been muddled up. Each input is with its own output.
Copy and complete this table with the missing input/output pairs.

Function machine	Correct input/output pairs	Muddled input/output pairs
×5 +3		$5 \longrightarrow 28$ $3 \longrightarrow 30$
		$9 \longrightarrow 60$ $7 \longrightarrow 38$
+3 ×5		$-1 \longrightarrow 10$ $-3 \longrightarrow -12$
		$0 \longrightarrow 3$ $0 \longrightarrow 15$

A6 HW3 Further equations

1 For each of these 'think of a number' problems

  i draw a function machine
  ii draw the inverse function machine
  iii use your inverse machine to work out the number.

  a I think of a number. I times it by 4 then add 6.
    The answer is 18.
  b I think of a number. I triple it and subtract 4.
    The answer is 14.
  c I think of a number. I multiply it by 5 and add −1.
    The answer is 39.
  d I think of a number. I divide it by 4 and subtract 5.
    The answer is 3.

2 Solve these equations.

  a $4n - 6 = 34$ **b** $2s + 4 = 14$ **c** $4f + 3 = -17$
  d $10 = 3x + 1$ **e** $20 - 5x = -5$ **f** $7d + 4 = -10$

3 Solve these equations using the balance method.
Check your answers by substitution.

  a $5n + 2 = 27$ **b** $30 - 2x = 28$ **c** $6d + 3 = -9$
  d $10p - 2 = 3$ **e** $8q - 3 = -11$ **f** $22 - 3e = 13$

4 The formula for the volume, V, of a cuboid is
$$V = lwh$$
where l = length, w = width and h = height.

  a Find the volume of a cuboid when $l = 6$ cm, $w = 5$ cm and
    $h = 2$ cm.
  b Which of the following cuboids will give the largest
    volume?

    i $l = 6$ cm, $w = 4$ cm, $h = 3$ cm
    ii $l = 7$ cm, $w = 4$ cm, $h = 2$ cm
    iii $l = 9$ cm, $w = 2$ cm, $h = 5$ cm

1 Whilst doing a science experiment, Sam is told to use the equation $v = 9.81t - 5.27$ to work out the value of v.

   a Work out the value of v when $t = 12$.

   b Work out the value of v when $t = 20$.

2 Solve these equations.

   a $6b + 2 = 20$

   b $30 = 5x - 5$

   c $\frac{x}{2} - 7 = 8$

   d $50 - 4x = 30$

3 The diagram shows a rectangle with side lengths $x + 4$ and 6.

   a Write an expression for the perimeter of the rectangle.

   b If the perimeter of the rectangle is 30 cm, find the value of x.

4 Write equations for these 'think of a number' problems and solve each equation.

   a I think of a number, multiply it by 8 and add 4. The answer is 44.

   b I think of a number, multiply it by 7 and add 6. The answer is 27.

   c I think of a number, multiply it by 9 and add −5. The answer is 49.

1 Solve these equations

   a $6x + 5 = 41$ **b** $10 - 2x = 4$ **c** $\frac{4}{x} + 1 = 5$

   d $20 - x = 22$ **e** $-2x = -30$ **f** $3x + 9 = -6$

2 a Plot and join the points
     $(-2, 0)$ $(-2, 2)$ $(1, 1)$
     on a copy of this grid.
  b Write down the
     mathematical
     name of this shape.
  c Rotate the shape
     through 90°
     anticlockwise
     about $(1, 1)$.

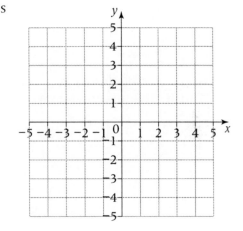

3 Hitesh invests £500 in a bank account.
  The bank pays 3% interest per year.

  a Work out how much interest Hitesh earns in one year.

  b The interest is added to the account.
     How much will Hitesh have in his account at the end
     of 1 year?

1 Make accurate drawings of these triangles.
Measure the unknown lengths in each triangle.

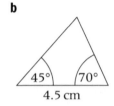

a

b

c
8 cm
67°
6 cm
45°   70°
4.5 cm
125°
5 cm

2 Make accurate drawings of these triangles.
Measure the unknown lengths in each triangle.

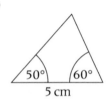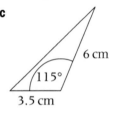

a

b

c
6 cm
47°
4 cm
50°   60°
5 cm
115°
3.5 cm

3 Make an accurate drawing of this trapezium.

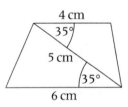

4 cm
35°
5 cm
35°
6 cm

4 a Make an accurate drawing of the sketch.

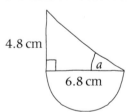

4.8 cm

6.8 cm

a

b On your drawing, measure the size of the angle marked *a*.

1 Find the bearings of all the places on the island from the treasure.

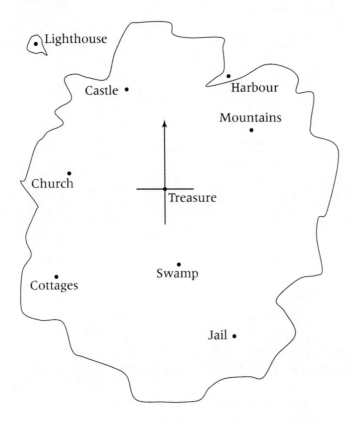

2 If the scale of the map is

1 cm represents $\frac{1}{2}$ km,

work out the distance of each place from the treasure.

3 If the scale of the map is

1 cm represents 250 metres,

work out the distance of each place from the treasure.

1

Stoke-on-Trent

Nottingham

N

Leicester

Birmingham

Coventry

Northampton

The diagram is part of a map showing the position of several towns.
Measure and write the bearing of

a Birmingham from Stoke-on-Trent
b Leicester from Coventry
c Leicester from Nottingham
d Birmingham from Northampton.

2 Make accurate drawings of these triangles.

a

b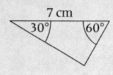

3 Construct an accurate scale drawing of this triangle.
Use a scale of 1 cm represents 2 cm.

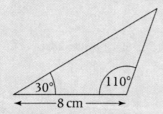

1 The diagram is part of a map showing the position of several places.

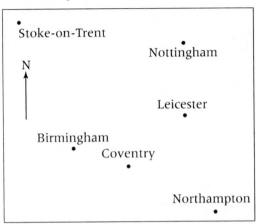

Measure and write down the bearing of

a Coventry from Birmingham
b Northampton from Coventry
c Stoke-on-Trent from Nottingham
d Leicester from Northampton.

2 Measure **a** the diameter   **b** the radius of this circle.

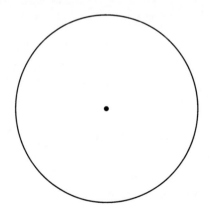

3 Draw a circle with diameter 8 cm.

1 a Find all the pairs of shapes that are congruent.
   b Name each shape.

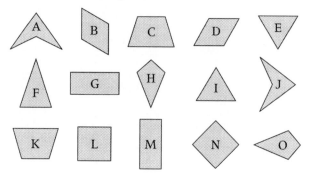

2 Find the pair of congruent triangles.

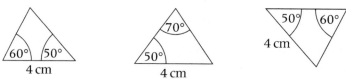

3 a Copy and complete this table.

Shape name and number of sides	Shape	Number of triangles	Sum of interior angles
Triangle 3 sides		1	$1 \times 180° = 180°$
Q _____ 4 sides		2	$2 \times 180° = 360°$
P _____ 5 sides		?	$? \times 180° = ?$
H _____ 6 sides		?	$? \times 180° = ?$

   b Extend the table up to an eight sided shape.

1 Find the circumference of these circles. Use $\pi = 3$.

Hint: Remember $C = 2\pi r$

a

5 cm

b

2 cm

c
3 cm

2 Work out the circumference of these circles. Use $\pi = 3.14$

   a Diameter = 4 cm   **b** Diameter = 6.5 cm
   c Radius = 3 cm   **d** Radius = 2.5 cm

3 Calculate the diameter of these circles. Use $\pi = 3$

   a Circumference = 36 cm
   b Circumference = 24 cm
   c Circumference = 10.5 cm

4 Calculate the area of these circles.
State the units of your answers.

a

radius = 6 cm

b

radius = 9 m

c

diameter = 36 mm

5 A circular trampoline has diameter 7 metres.
Calculate the area of the trampoline.

6 Mary is fencing off a circular area for her chickens.
She has marked out a circle with radius 3 metres.
How much fencing does she need to go all around the edge
of this circle?

1 Arrange 6 identical squares into as many nets of a cube as possible. One is done for you.

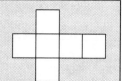

2 a Draw the nets of these cuboids.
 b Work out **i** the surface area
  ii the volume of each cuboid.

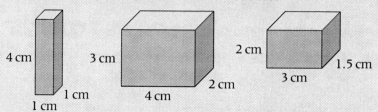

3 The diagram represents a tank in the shape of a cuboid.
The tank has a base.
It does not have a top.
The length of the tank is 2.6 m.
The width of the tank is 1.8 m.
The height of the tank is 3.5 m.

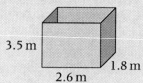

 a The outside of the tank is going to be painted. Work out in m² the amount of paint that will be needed.
 b The tank is going to be filled with sand. Work out the maximum amount of sand that the tank will hold.

4 Match up the congruent shapes.

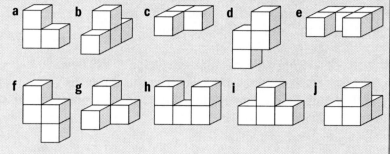

GCSE Formulae

In your AQA GCSE examinations you will be given a formula sheet like the panel below.

Area of a trapezium = $\frac{1}{2}(a + b)h$

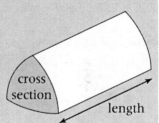

Volume of prism
= area of cross section × length

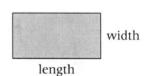

Here are some other formulae
that you should learn.
These will not be given in your exams.

Area of a rectangle = length × width

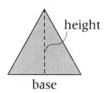

Area of a triangle = $\frac{1}{2}$ × base × height

Area of a parallelogram
= base × perpendicular height

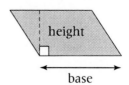

Area of a circle $= \pi r^2$

Circumference of a circle $= \pi d = 2\pi r$

Volume of a cuboid
$=$ length $\times$ width $\times$ height

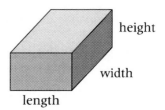

Volume of a cylinder
$=$ area of circle $\times$ length

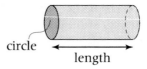

Pythagoras' theorem states,
For any right-angled triangle, $c^2 = a^2 + b^2$
where c is the hypotenuse.

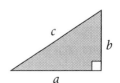